T0231092

VIKING DNA

The Wirral and West Lancashire
Project

by

Stephen Harding,
Mark Jobling,
and
Turi King

Countyvise Limited, UK

CRC Press
Taylor & Francis Group
Boca Raton London New York

CRC Press is an imprint of the
Taylor & Francis Group, an **informa** business

Front cover illustration: Vikings arrive in Wirral © Chris Collingwood, http://www.chriscollingwood.co.uk

Back cover illustration: Steve Harding, Turi King and Mark Jobling brandishing a mouth swab. Courtesy of Terje Størksen, Haugesunds Avis.

First Published 2010 by Countyvise Limited, 14 Appin Road, Birkenhead, Merseyside, CH41 9HH, UK.

Co-published with Nottingham University Press, Manor Farm, Church Lane, Thrumpton, Nottingham, NG11 0AX, UK.

The right of Stephen Harding, Mark Jobling and Turi King to be identified as the authors of this work has been asserted by them in accordance with the Copyright, Design and Patents Act 1988.

CRC Press
Taylor & Francis Group
6000 Broken Sound Parkway NW, Suite 300
Boca Raton, FL 33487-2742

© 2010 by Stephen Harding, Mark Jobling and Turi King.
CRC Press is an imprint of Taylor & Francis Group, an Informa business

No claim to original U.S. Government works

Printed on acid-free paper
Version Date: 20150217

International Standard Book Number-13: 978-1-4665-9085-4 (Paperback)

This book contains information obtained from authentic and highly regarded sources. Reasonable efforts have been made to publish reliable data and information, but the author and publisher cannot assume responsibility for the validity of all materials or the consequences of their use. The authors and publishers have attempted to trace the copyright holders of all material reproduced in this publication and apologize to copyright holders if permission to publish in this form has not been obtained. If any copyright material has not been acknowledged please write and let us know so we may rectify in any future reprint.

Except as permitted under U.S. Copyright Law, no part of this book may be reprinted, reproduced, transmitted, or utilized in any form by any electronic, mechanical, or other means, now known or hereafter invented, including photocopying, microfilming, and recording, or in any information storage or retrieval system, without written permission from the publishers.

For permission to photocopy or use material electronically from this work, please access www.copyright.com (http://www.copyright.com/) or contact the Copyright Clearance Center, Inc. (CCC), 222 Rosewood Drive, Danvers, MA 01923, 978-750-8400. CCC is a not-for-profit organization that provides licenses and registration for a variety of users. For organizations that have been granted a photocopy license by the CCC, a separate system of payment has been arranged.

Trademark Notice: Product or corporate names may be trademarks or registered trademarks, and are used only for identification and explanation without intent to infringe.

Visit the Taylor & Francis Web site at
http://www.taylorandfrancis.com

and the CRC Press Web site at
http://www.crcpress.com

"Connections between genes and surnames represent an exciting area in current Viking research, contributing important and new information to the picture painted by history, place-names and archaeology. This accessible account of a major scientific project focused on north-west England will stimulate many new ideas about the period, and the results described here give fresh impetus to the ways in which we view the Viking legacy in today's population."
David Griffiths, Reader in Archaeology, University of Oxford.

"I am happy to have been associated, in a very small way, with the innovative and exciting research project described in this book. Thanks largely to the energy and enthusiasm of Professor Stephen Harding, the Wirral and West Lancashire project has rekindled the interest of the people of Merseyside in their rich and varied heritage, revealing the part played in that heritage by the population movements of the Viking Age. This work and other recent discoveries in the area have suggested many more interesting questions which will keep Viking researchers (including myself) busy for some years to come."
Judith Jesch, Professor of Viking Studies, University of Nottingham.

"I warmly welcome this new book by Stephen Harding, Mark Jobling and Turi King which ably and fascinatingly demonstrates the extent of how the legacy of the Vikings in Wirral (and our close neighbour West Lancashire!) has been carried right through to todays world through the genes of local people. Professor Harding, through a number of highly readable books in recent years, has done much to reveal the true nature of the Viking presence in Wirral from the late 9th century and the evidence for this through for example local place names, historical sources and archaeology. In this latest publication an entirely new dimension is researched literally through the descendants of early Viking settlers within the present day Wirralian population. Wirral is justly proud of its Viking heritage and recognises the role, impact and legacy of the Vikings in the evolution of the Wirral Peninsula over the last 1100 years. That legacy echoes through our classrooms and through the interest of both local people and our visitors."

Howard Mortimer, Head of Special Initiatives, Wirral Council.

"West Lancashire Heritage Association has been delighted to have been associated with this project, helping to facilitate the formation of the required sample of male volunteers with old West Lancashire roots, and then the sampling, the communication and meetings of the group and the final publication of the West Lancashire results. For this aspect of the research over 200 descendents, their relations and dignatories from West Lancashire and our neighbours from Wirral heard at a special event held on "Lancashire Day" - 27th November - the news of just how strong the DNA links of the people from this area are with their Viking ancestors. In fact many of the people present showed they also had the special medical condition involving the hand known as Dupuytrens Contracture which has also been closely associated with the Vikings! The research was conducted to the highest of standards and has done a great deal to elevate the profile of West Lancashire and our Association."

Patrick Waite, Chairman, West Lancashire Heritage Association.

Contents

Foreword
by Michael Wood

The Wirral is one of the most distinctive and enigmatic landscapes in Britain. Its wedge shaped peninsula bounded by the Dee, the Mersey and the Irish Sea, is situated between England and Wales and, or so it seemed to some early travellers, not quite belonging to either. Facing out towards Ireland across 'our sea, mare nostrum', as St Patrick called it in the fifth century, its anchorages and trading shores have seen human migration, commerce and invasion for millennia. In the early 19th century when a shift in tidal currents began to move the sea front dunes on the tip of the peninsula at Meols, thousands of ancient artefacts from the Bronze Age to the 15th century were exposed, revealing that this wide windswept trading shore had been used from prehistory. The finds included scores of Roman brooches and coins, and even a silver tetradrachm minted in Syria in the Hellenistic age. Byzantine coins and a 6th century pilgrims flasks from the shrine of St Menas near Alexandria in Egypt unexpectedly showed that the Wirral's links with the Mediterranean had continued into the Dark Ages. From the Viking era masses of small finds – strap ends, mounts, and coins including over twenty Anglo-Saxon silver pennies - showed Meols was still the main trading place in the Irish Sea. So around 900, when the narrative of this fascinating book is centred, the wide trading shore at Meols was still active, the main entry point into northern England from Ireland.

And in neighbouring West Lancashire another popular entry point was the Ribble, and here along its banks one of the most outstanding collections of Viking Treasure was found, also in the 19th century – the magnificient Cuerdale Hoard, not far from Preston. And nearby can be found another Meols – North Meols, and not far away Ravenmeols and a former Argarmeols.

The name Meols like many in the Wirral and West Lancashire is Viking, meaning 'The Sand Dunes': from the Wirral coast it was a day and a half with a good wind to the thriving Viking colony in Dublin; a couple of hours ride to Chester, the old Roman city of Deva soon to become a major Anglo-Saxon emporium after its ancient walls were restored by Æthelflæd

'Lady of the Mercians' in 907. The rapid growth in the wealth of Chester, which is also the background to this tale, must have come from its position at the head of Wirral, as is suggested by the appearance of at least twenty five moneyers working for the English King Athelstan in Chester in the 930s - more than any other city in England. The Chester Esplanade hoard of 550 coins and 150 pieces of hacksilver deposited in the 960s is another of a number of rich finds from this time which underline the fast increasing wealth of this corner of the North West. And to judge by their names, the Chester moneyers came from many races: Norse, Irish, Welsh and English, a typical mix of the region at this time. Viking Chester, Wirral and West Lancashire then were going places: a precursor of the medieval and modern cities which sprang up on the shores of the Irish Sea, and whose destiny would be taken over eventually by the great port of empire - Liverpool, the melting pot of the modern North West.

The tides of history then run long and deep around Wirral and West Lancashire… People have always been drawn here from the coast of Ireland, from Wales, and from the islands and archipelagos of North Britain, across what one might call the free trade zone of the Irish Sea. The Wirral peninsula is only two hundred miles from London, but till relatively recently it inhabited another world. Turn the map round 90 degrees, so that Dublin is at the bottom, the Isle of Man in the centre, and the Dee, Mersey and Ribble at the top. Then imagine a different mental map. You will instantly see more clearly that the Irish Sea is really like an inland sea, and you will see why through history the shores of Ireland, North Wales and south western Scotland have had more contact with each other, and perhaps even more in common with each other, than with Southern England. In the Dark Ages, the age of saints and seaways, it was the connections across this sea that shaped local peoples lives. They still do today of course: to take only the most obvious example, that of local names, of say footballers - Carragher, Rooney, and Gerrard; of comedians like Askey, Tarbuck and O'Connor; or of musicians – Harrison, Lennon, McCartney and Starkey: their names Norse, Irish, Gaelic, or English in origin, speak of those deep connections across and around the Irish Sea: a circulation of people in the modern age that was already there in the deep past.

That is the background to the literally fantastic story told in this book. The Norse settlement of the Wirral began in 902 with another wave of migrants, some from Ireland, some from Man and the Hebrides perhaps, many no doubt of already mixed Norse-Irish ancestry. After the refounding of the old Roman city of Chester in 907 there was fighting between them and the rulers of English Mercia, but this was followed by a treaty and by what may have been a formal ceding of the northern part of the peninsula to a self governing Norse oligarchy. This series of events has left its mark till today in the Viking place names of the Wirral from Birkenhead to Tranmere and West Kirby, and over the Mersey in such well known Liverpool and Lancashire names as Toxteth, Croxteth, Kirkby, West Derby, Ormskirk, Skelmersdale and Scarisbrick which are also of Viking origin. Wirral and West Lancashire boast the only places in England with definite examples of the name Thingwall – the Viking place of Assembly or Self-Government. But as this remarkable book tells, we now know that the Norse settlement of the Wirral and West Lancashire in the early 900s left its mark in another way, as is revealed by exciting scientific discoveries which have traced its genetic impact in the old families of the Wirral and West Lancashire right down to today. This book is based on fascinating documentary research, but most intriguingly on a new DNA examination of the population of the North West. Spared many of the ravages of history the communities of the Wirral for a long time preserved very old patterns of social organisation and inheritance, customs of marriage and nomenclature. Here cutting edge science has been helped by the old fashioned magic of the parchment trail: medieval rent books for example show that as late as the 15th century Scandinavian customs were still maintained in the Wirral by people still bearing Viking names, such as Agnes 'Hondesdoghter' and Mabilla 'Raynaldesdoghter'. To this story the book adds amazing detail by tracing the descent of such people in living Wirral and West Lancashire families; a truly remarkable piece of historical detective work which involves a brilliant combination of landscape history, philology, science and textual criticism, and intensive local research.

It is hard to imagine many places in the UK (given the huge population changes everywhere since the early nineteenth century) where the accident of history and the survival of sources allows us such a close up view. The result as the reader will see in these pages is nothing short of haunting.

For as the best history always should, the tale also involves identity, local feeling, the life lived; it suggests how in a mysterious way even the deep past still lives on in us. It also shows how gripping and informative local history can be, but how it also can vividly illuminate the big picture. It is a model applicable to every community in the UK: a perfect example of what one hopes will become the new history. Not one delivered from on high but developed at grass roots in conjunction with the people themselves, using their archaeology history and landscape, their family histories documents and memories -and even their DNA. In its pages we sense the givenness of a past going back thousands of years, but vividly brought to life when a mixed band of Norse, Irish and Hebridean adventurers put down roots along the shores of the Wirral and West Lancashire at 'Crane's sandbank' at 'the Dunes' on 'Birch Tree headland' and 'Forni's settlement'.

29th October, 2010

Preface

In 2002 a rigorous genetic survey of the Wirral peninsula and West Lancashire – two of the key areas of North West England known from place-name, archaeological and historical evidence to have had clear Viking presence - was commenced. The study was based on DNA from the male-specific Y chromosome, which is passed down from generation to generation along the paternal line with little or no change – providing an ideal probe for population ancestry. It was led by a team of scientists from the Universities of Leicester and Nottingham, together with University College London, and helped by historical and linguistic experts and the local communities. We used old surnames that were present in Wirral and West Lancashire prior to 1600 as a guide to the recruitment of volunteers. The results were presented at a meeting in November 2007 in Knowsley, West Lancashire and published in the prestigious scientific journal *Molecular Biology and Evolution* in February 2008. At the request of many people we have produced this comprehensive report describing as clearly as we can for the non-expert the background, results, conclusions and perspectives coming out of this research which strongly supports the idea that the northwest of England was once heavily populated by Scandinavians.

This research would not have been possible without the generous support of many people and organisations. We thank all DNA donors and also BBC Radio Merseyside, BBC Radio Lancashire, the Daily Post, Wirral News, Wirral Globe, and Ormskirk Advertiser for assistance in the recruitment of volunteers. We thank fellow co-authors of the *Molecular Biology and Evolution* paper: Georgina Bowden, Patricia Balaresque and Andrew Lee at the University of Leicester, Ziff Hansen, Giles Pergl-Wilson, Emma Hurley and Professor Judith Jesch at the University of Nottingham; Professor Mark Thomas and Abigail Jones at University College London, Stephen Roberts of The Queen Catherine School, Kendal Cumbria - author of *A History of Wirral* - and Patrick Waite of the West Lancashire Heritage Association. Dr. Chris Tyler-Smith (University of Cambridge) and Professor Agnar Helgason (University of Reykjavik) helped with the admixture analysis and Dr. Neil Bradman (University College London) also contributed to some of the haplotyping. We thank all the people of the West Lancashire Heritage

Association (particularly Jennifer Waite, Angela Tregent, Jim Vose and the late Ruth Vose) and also Dr. David Favager of Wirral Grammar School and all the dentists who helped with the recruitment of some of the volunteers for the modern control samples.

The research was supported by a special Watson-Crick DNA anniversary award grant from the U.K. Biotechnology and Biological Sciences Research Council (to Steve Harding, Mark Jobling and Judith Jesch) to commemorate the 50th anniversary of the discovery of the DNA double helix by Crick and Watson in 1953. It was also supported by the Wellcome Trust as a Senior Fellowship (to Mark Jobling) and a Prize Studentship (to Turi King), and the Melford Charitable Trust.

Steve Harding
Mark Jobling
Turi King

Nottingham and Leicester, October 2010

This book is dedicated to the memories of
John Emmerson,
Publisher,
1936 – 2010
and
J. Michael Creeth,
Scientist,
1924 - 2010

Chapter 1
INTRODUCTION

The Vikings were a fierce seafaring people originating from Scandinavia who long ago raided and traded with the British Isles. There is also evidence of many establishing themselves and forming lasting settlements. For some it was a case of an attempted military conquest, as manifested in the well chronicled battles of the *Danes* with King Alfred the Great and the establishment of the Danelaw – a separate, self-governing Danish territory - in northern and eastern England. For others it was simply a somewhat more peaceful desire to explore and find bases where they could trade or to find lands where they could make a living by farming and fishing, but records suggest even these peoples were prepared to use violent means when it suited.

The historical records, place-name evidence and archaeological finds all point to the arrival and presence of Vikings of primarily Norwegian origin – *Norsemen* – in North West England over a millennium ago[1]. These were Vikings who had been operating around the Irish Sea and when they entered the northwest coastal regions of England would have brought Danish and Irish settlers with them. The migration was particularly great after a huge exodus in the year 902 from attempted settlements in Ireland and one of the greatest ever finds of Viking treasure – the Cuerdale hoard on the banks of the River Ribble[2] – together with other hoards found in and near the city of Chester, such as the recently described Huxley hoard[3], all date from around this period.

---

1. Cavill, P., Harding, S.E. and Jesch, J. (2000) *Wirral and its Viking Heritage*. English Place-Name Society, Nottingham, UK.; Harding, S.E. (2000) *Ingimund's Saga. Norwegian Wirral*. Countyvise, 2nd edition, 2006, Birkenhead, UK.; Harding, S.E. (2002) *Viking Mersey: Scandinavian Wirral, West Lancashire and Chester.* Countyvise, Birkenhead, UK.; Griffiths, D. (2010) *Vikings of the Irish Sea*. History Press, Stroud UK.
2. Graham-Campbell J.A., ed. (1992) *Viking Treasure from the North West*. National Museums and Galleries on Merseyside Occasional Papers, Liverpool, UK.
3. Graham Campbell J.A. and Philpott, R.A. eds. (2010) *The Huxley Viking Hoard. Scandinavian Settlement in the North West*, National Museums of Liverpool, UK.

The coastal regions from the River Dee to the River Solway have a wealth of place-names of Scandinavian or Irish-Scandinavian origin. In England, however, only Wirral and West Lancashire possess definite examples of the name Thingwall (from Old Norse þing-völlr, meaning assembly field), indicating settlements of sufficient density and autonomy to warrant their own parliament (the nearest related place-names are Tynwald in the Isle of Man and Tinwald in South West Scotland). As a starting point for a detailed study on Viking ancestry in northern England as a whole, the Wirral

FIGURE 1-1. *Cavendish Laboratory Annex, Cambridge 1953. Francis H. Crick and James D. Watson demonstrate their double-helical model for the structure of DNA[4] which won them the Nobel Prize for Medicine with M.H.F. Wilkins of King's College London[5]. 50 years on an anniversary fund was set up to support important areas of DNA research and the Wirral and West Lancashire project was one of those areas supported.*

4. Photo by Antony Barrington Brown and reproduced with permission from the Science Photo Library.
5. See Watson, J.D. (1970) *The Double Helix: A Personal Account of the Discovery of the Structure of DNA*, Penguin, London, UK.

peninsula and West Lancashire therefore presented a particularly attractive area for study.

But what of evidence for ancestry from these Viking settlers in the genes and blood of people living in Wirral and West Lancashire today? The year 2002 provided the ideal opportunity for the start of a programme to probe this ancestry. Not only did this year mark the 1100th anniversary of the expulsion of the Norsemen from Ireland but also marked the eve of the 50th anniversary of arguably the greatest scientific discovery of the 20th Century – the double-helical structure for DNA - by two young scientists at the University of Cambridge, James D. Watson and Francis H. Crick (Figure 1-1) a finding published in the scientific journal *Nature*[6]. To commemorate this the Biotechnology and Biological Sciences Research Council of the United Kingdom provided support for a genetic study of the Wirral and West Lancashire to see if there was a strong Viking legacy remaining in the DNA of people from old Wirral and West Lancashire families still living there today. This research was also supported by the Wellcome Trust. The survey was completed in 2007, after exhaustive work by a team of 13 researchers, and the results were presented at a meeting attended by 190 invited people including participants in the survey and dignitaries from Wirral and West Lancashire (Figure 1-2). This meeting was held at Knowsley Hall in South West Lancashire in November 2007 and was followed by an open presentation at St. Bridget's Church West Kirby, Wirral attended by nearly 300 people. The results appeared in publication form in 2008 in the leading scientific journal *Molecular Biology and Evolution* (Figure 1-3), confirming what was suspected previously from the wealth of historical, place-name and archaeological evidence, such as the artefacts shown in Figures 1-4 and 1-5.

---

6. Watson, J. D. and Crick F.H.C. (1953) A structure for deoxyribose nucleic acid, *Nature*, volume 171, pages 737-738.

FIGURE 1-2. *Steve Harding, Judith Jesch and Mark Jobling with Viking re-enactor Kevin Taylor at the event at Knowsley Hall, November 2007*[7].

7. Photo courtesy of Trinity Mirror Copyright and the Liverpool Echo.

Excavating Past Population Structures by Surname-Based Sampling: The Genetic Legacy of the Vikings in Northwest England

Georgina R. Bowden,* Patricia Balaresque,* Turi E. King,* Ziff Hansen,† Andrew C. Lee,*[1] Giles Pergl-Wilson,† Emma Hurley,† Stephen J. Roberts,‡ Patrick Waite,§ Judith Jesch,‖ Abigail L. Jones,¶ Mark G. Thomas,# Stephen E. Harding,† and Mark A. Jobling*

*Department of Genetics, University of Leicester, Leicester, United Kingdom; †National Centre for Macromolecular Hydrodynamics, University of Nottingham, Sutton Bonington Campus, Loughborough, United Kingdom; ‡The Queen Katherine School, Kendal, Cumbria, United Kingdom; §West Lancashire Heritage Association, Ormskirk, United Kingdom; ‖School of English Studies, University of Nottingham, University Park, Nottingham, United Kingdom; ¶The Centre for Genetic Anthropology, Department of Biology, University College London, London, United Kingdom; and #Department of Biology, University College London, London, United Kingdom

The genetic structures of past human populations are obscured by recent migrations and expansions and have been observed only indirectly by inference from modern samples. However, the unique link between a heritable cultural marker, the patrilineal surname, and a genetic marker, the Y chromosome, provides a means to target sets of modern individuals that might resemble populations at the time of surname establishment. As a test case, we studied samples from the Wirral Peninsula and West Lancashire, in northwest England. Place-names and archaeology show clear evidence of a past Viking presence, but heavy immigration and population growth since the industrial revolution are likely to have weakened the genetic signal of a 1,000-year-old Scandinavian contribution. Samples ascertained on the basis of 2 generations of residence were compared with independent samples based on known ancestry in the region plus the possession of a surname known from historical records to have been present there in medieval times. The Y-chromosomal haplotypes of these 2 sets of samples are significantly different, and in admixture analyses, the surname-ascertained samples show markedly greater Scandinavian ancestry proportions, supporting the idea that northwest England was once heavily populated by Scandinavian settlers. The method of historical surname-based ascertainment promises to allow investigation of the influence of migration and drift over the last few centuries in changing the population structure of Britain and will have general utility in other regions where surnames are patrilineal and suitable historical records survive.

FIGURE 1-3. *Publication in Molecular Biology and Evolution, February 2008[8]. Reproduced, courtesy of Oxford University Press.*

8. Bowden, G.R., Balaresque, P., King, T.E., Hansen, Z., Lee, A.C., Pergl-Wilson, G., Hurley, E., Roberts, S.J., Waite, P., Jesch, J., Jones, A.L., Thomas, M.G., Harding, S.E. and Jobling, M.A. (2008) Excavating past population structures by surname-based sampling: the genetic legacy of the Vikings in North West England. *Molecular Biology and Evolution*, volume 25, pages 301-309.

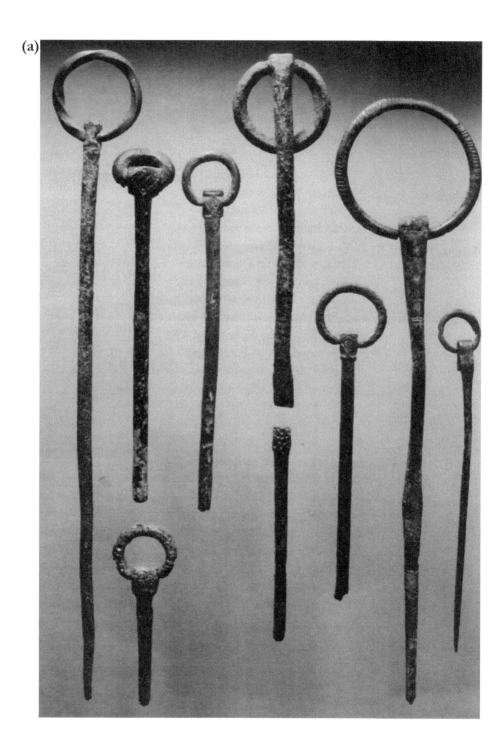

(b)

FIGURE 1-4. *Some of the great abundance of Viking treasure of North West England: (a) Ringed dress pins from the former Viking seaport of Meols, Wirral (9th-12th Century). Courtesy of David Griffiths, University of Oxford. (b) Part of the largest hoard of Viking treasure ever found – Cuerdale, West Lancashire (~AD 905). Courtesy of David Flower © National Museums Liverpool (World Museum).*

FIGURE 1-5. *Hogback Viking tombstone (10th Century), one of two found in Wirral. Courtesy of Jennifer Whalley and Peter Crawford. Photo: Phil Hirst.*

The research team were as follows:

Mark Jobling, Professor of Genetics at the University of Leicester and expert on the link between surnames and the human Y chromosome. Mark featured in BBC's *Motherland* programme.

Steve Harding, Professor of Applied Biochemistry at the University of Nottingham, Director of the Nottingham Centre for Macromolecular Hydrodynamics and expert on the Vikings in the area of his origin – Wirral.

Turi King and **Patricia Balaresque**, Post-doctoral research associates at the University of Leicester. Turi, like Mark, is an expert on the link between surnames and the human Y-chromosome. Patricia is an expert on the evolutionary dynamics of the sex chromosomes.

Judith Jesch, Professor of Viking Studies at the University of Nottingham and co-author with Steve Harding and Paul Cavill of *Wirral and its Viking Heritage* (English Place Names Society, 2000).

Stephen Roberts, Queens School, Lancaster. Author of the books *A History of Wirral* (Phillimore, 2002) and *Hoylake and Meols Past* (Phillimore, 1992) and an expert on the origin of surnames in the north west of England.

Patrick Waite, Chairman of the West Lancashire Heritage Association, who organised volunteer recruitment.

Mark Thomas, Professor of Population Genetics and **Abigail Jones** (Researcher) at University College London, and **Giles Pergl-Wilson** and **Emma Hurley** (MSc students at the University of Nottingham) were responsible for providing the data representing the modern populations.

Georgina Bowden and **Ziff Hansen** (technical workers at the Universities of Leicester and Nottingham) carried out most of the laboratory work.

In order to properly assess the extent of Viking ancestry it was necessary to deal with the complication of post Viking migration into Wirral and West Lancashire particularly following the Industrial Revolution and the development of Liverpool as a port. Since 1800, for example, Wirral has experienced a population increase from about 8000 to around 350,000 – this is approximately 6 times the average national increase for England. Fortunately the existence of extensive lists of surnames of people who were resident in the region prior to 1600 – from tax records, alehouse records, criminal records and an extensive list of those contributing towards the stipend of a priest - helped guide the recruitment of volunteers.

This book is about this survey – the background of history, place-name evidence and archaeological evidence, the method of linking old genes with modern geography - focusing on men from old families present in the area prior to the Industrial revolution - the results and their importance, and also how this survey has led as a platform for a further investigation extending across the whole of northern England to see how far these Vikings entering from the Irish Sea had penetrated. Another huge hoard of Viking treasure found near Harrogate, plus the presence of Irish-Norse "hogback" tombstones found in North West England but also well into North Yorkshire suggests that this penetration was considerable. It is also about the people who took part. We start, however, with the historical evidence and the remarkable story of the arrival of one group of Vikings in Wirral in the early part of the 10th Century.

Chapter 2
BACKGROUND: HISTORICAL, PLACE-NAME AND ARCHAEOLOGICAL EVIDENCE OF VIKING SETTLEMENT

The first piece of evidence for Viking presence in Wirral and West Lancashire comes from historical records – not the Anglo-Saxon chronicles, or at least not directly, but from Irish and Welsh annals describing the expulsion of Vikings from Dublin followed by an influx into Wirral of "mass migration" proportions[9] (Figure 2-1). But the Viking story starts a lot earlier.

FIGURE 2-1. *Vikings led by Ingimund arrive in Wirral sometime in or after AD 902*[10].

9. Wainwright, F.T. (1975) *Scandinavian England: Collected Papers.* Ed. Finberg, H.P.R., Phillimore, Chichester.

10. Painting by Chris Collingwood. http://www.chriscollingwood.co.uk/

Harald Hårfagre

The cause of what we now know as the Viking Age is still a matter of debate but the consensus of opinion appears to be that it was due to a combination of several factors[11]. One of these was a great thirst for adventure of young Scandinavian men: one of the famous sagas, for example, reports how the young Icelander Egil Skallagrimsson dreamt of becoming a Viking[12]. This desire was probably reinforced by an overcrowding of many of those parts of Scandinavia able to sustain a population – an increasing scarcity of good farming land and coastal areas to fish in. The second factor was the availability of the means to achieve this – the development of the longboat – a superb piece of craftsmanship emanating from Scandinavia. The longboat evolved from a clinker-built craft without a sail – an early example, dated to around AD 400, was found at Nydam on the Jutland peninsula[13]. Once a strong keel – which held the sail – became available, the potential to explore lands afar became a reality. Several examples have been discovered, the most notable being the Oseberg and Gokstad ships now displayed at the Oslo Ship Museum in Bygdøy. Another important factor behind the Viking Age was the changing political climate in Norway and in particular the emergence of a great hulking figure from the latter part of the 9th Century who united the many Norwegian kingdoms under one control – his own. That man was Harald Hårfagre (Figures 2-2, 2-3). According to sources such as *Heimskringla*[14] written by Snorri Sturluson sometime between 1223 and 1225 and *Egil's Saga* – also considered by scholars to have been written by the same author - Harald Hårfagre was responsible for uniting the several independent kingdoms under one throne:

Haraldr, son Hálfdanar svarta, hafði tekit arf eptir föður sinn; hann hafði þess heit strength, at láta eigi skera hár sitt né kemba, fyrr en hann væri einvaldskonungr yfir Noregi,

11. Jones, G. (2001) *A History of the Vikings* (2nd edition). Oxford University Press.
12. Sturluson, S. Egil's Saga, in Thorsson, Ö. (editor) and Scudder, B. (translator) (2001) *The Sagas of the Icelanders*, Penguin Books, New York.
13. Magnusson, M. (2000) *The Vikings*. Tempus Publishing, Stroud, page 18.
14. Sturluson, S. in Foote, P. (editor) and Laing, P. (translator) (1961) *Heimskringla*, Everyman's Library, J.M. Dent & Sons, London.

which translates as "Harald, son of Hálfdan the Black, had succeeded his father; he had made a solemn vow neither to cut nor comb his hair until he was king of all Norway". According to the saga unification of Norway was a requirement of the Hordaland woman *Gyda* before she would agree to marry him. The subduing of the Norwegian kingdoms Harald subsequently achieved by a series of battles and treaties, culminating in one great battle at Hafrsfjordur (now Hafrsfjord, just west of Stavanger) which took place around the year AD 890: according to *Heimskringla* : "after this battle King Harald met no opposition in Norway for all his worst enemies had fallen. But some, and they were a great number, fled out of the country and thereby great uninhabited districts were peopled".

FIGURE 2-2. *Model of Harald Hårfagre and his wife Gyda at the Nordvegen Historical Centre, Karmøy.*

FIGURE 2-3. *Reconstruction underway at Karmøy, Haugesund of Harald Hårfagre's ship Draken (The Dragon). This is a project led by Norwegian businessman Sigurd Aase in conjunction with Marit Synnøve Vea, Terje Andreassen and Harald Løvvik. The original vessel would have been operating in the northern seas about the same time as Ingimund's epic voyage in AD 902/3. Left to right: Harald Løvvik, Mark Jobling, Turi King and Terje.*

Dublin

Well before the battle of Hafrsfjord, Vikings were already marauding down the Irish Sea. In or around the year AD 841 they discovered an ideal base for their operations at an inlet along a river on the east coast of Ireland – this was the River Liffey and the inlet became known as the "Black Pool" – dyf-lin or Dublin. The inlet was ideal for mooring and repairing their ships and was also located near a monastery which the Vikings would have found easy pickings to plunder. Dublin became an area of intense Viking settlement particularly around the area known as Wood Quay where there have been extensive excavations of the former Viking town. Dublin provided the main base for their operations in the Irish Sea and also a major centre for trade. However, at the beginning of the 10th Century that all changed.

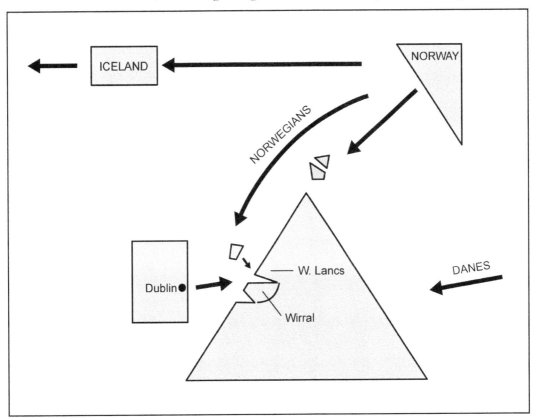

FIGURE 2-4. *The exodus from Norway. Drawing by John Harding.*

The following is an extract from ancient Irish annals known as the *Three Fragments* which report a huge battle in the year AD 902 and the resulting expulsion of the Vikings from Ireland leading to a mass exodus across the Irish Sea[15] (Figure 2-4). The annals tell of how they tried unsuccessfully to land and settle in Anglesey, how they were subsequently driven from there too, and then how their leader – a man called Ingimund (called Hingamund by the Irish) - approached the Queen of the Mercian English - Æthelflæd, daughter of Alfred the Great - asking for permission to settle in lands in the North West, and how she gave them permission to settle in "lands near Chester" – which scholars accept has to be Wirral.

THE STORY OF INGIMUND

(from "Fragmenta Tria Annalium Hiberniæ", Bibliothèque Royale, Brussels, MS. 5301-5320, fo. 33a-fo. 34b). Translation by the late Professor I.L. Foster, of Jesus College, Oxford.

We have related above, namely in the fourth year before us, of the expulsion of the Norse hosts from Ireland; through the fasting and praying of the holy man, namely Céle Dabhaill, for he was a saintly, devout man. The Norsemen, then, departed from Ireland as we have said and Hingamund was their leader, and where they went to was the island of Britain[16]. The king of Britain at this time was the son of Cadell, son of Rhodri. The men of Britain[17] assembled against them, and they were driven by force from the territories of the men of Britain.

Afterwards Hingamund came with his forces to Edelfrida queen of the Saxons, for her husband, that is Edelfrid, was that time in disease (let no one blame me although I have already mentioned the death of Edelfrid, and it was from the disease that Edelfrid died, but I did not wish to leave unwritten what the Norsemen did after going from Ireland). Now Hingamund was asking lands of the queen in which he would settle, and on which he would build huts and dwellings, for at this time he was weary of war. Then Edelfrida gave him lands near Chester, and he stayed there for a long time. The result of this was,

15. See for example Wainwright, F.T. (1975) *Scandinavian England: Collected Papers*. Ed. Finberg, H.P.R., Phillimore, Chichester, UK, and Harding, S.E. (2006) *Ingimund's Saga*. Countyvise Ltd., Birkenhead, UK.
16. Island of Britain = Anglesey.
17. Britain = Wales. The Vikings called Wales Bretland, "Land of the Brits".

when he saw the city full of wealth and the choice of land around it, he desired to possess them.

Afterwards Hingamund came to the leaders of the Norsemen and the Danes; he made a great complaint in their presence, and he said that they were not well off without good lands, and that it was right for them all to seize Chester and to possess it with its wealth and its lands. Many great battles and wars arose on account of that. This is what he said: "Let us beseech and implore them first, and if we do not get them willingly in this way let us contest them by force". All the leaders of the Norsemen and the Danes agreed to do this. Ingimund then came to his house, with an assembly following him. Though they made this council a secret, the queen came to know of it. Therefore the queen collected large forces around her in every direction, and the city of Chester was filled with her hosts.

The armies of the Danes and Norsemen assembled towards Chester and, since they did not get their consent by beseeching or supplication, they proclaimed battle on a certain day. On that day they came to attack the city; there was a large force with many freemen in the city awaiting them. When the forces who were in the city saw, from the wall of the city, the great armies of the Danes and Norsemen approaching them, they sent messengers to the king of the Saxons who was in a disease, and on the point of death at that time, to ask his advice and the advice of the queen. This was the advice they gave: to make battle near the city outside, and the gate of the city should be wide open, and to choose a body of horsemen, concealed on the inside, and those of the people of the city who should be stronger in the battle should flee back into the city as if in defeat, and when the greater number of the forces of the Norsemen came inside the gate of the city the force hidden yonder should close the gate after this band and not admit any more; capture those who came into the city and kill them all. This was all done accordingly, and complete slaughter was thus made of the Danes and Norsemen. Great, however, as was that slaughter, Norsemen did not abandon the city, for they were stubborn and vicious, but they all said that they would make many hurdles, and put posts into them, and pierce the wall under them. This was not delayed; the hurdles were made, and the forces were under them to pierce the wall, for they were eager to take the city to avenge their people.

Then the king (who was on the point of death) and the queen sent messengers to the Irishmen who were among the pagans (for there were many Irish among the pagans), to say to the Irishmen: "Life and health to you from the king of the Saxons, who is in disease, and from his queen, who has all authority over the Saxons, and they are certain that you are true and trusty friends to them. Therefore you should take their side; for they did not bestow any greater honour to a Saxon warrior or cleric than to each warrior and cleric who came to them from Ireland, because this inimical race of pagans is equally hostile to you also. It is right, then, for you, as you are trusty friends, to help them on this occasion." This was the same as if it was said to them: We have come from faithful friends of yours to address you so that you ask the Danes what token of lands and treasures they would give to those who would betray the city to them. If they accept this, to bring them to swear to a place where it will be easy to kill them; and when they will be swearing by their swords and by their shields, as is their custom, they will lay aside all their missile weapons. They all did accordingly, and they put away their arms. And the reason why the Irishmen did this to the Danes was because they were less friends to them than to the Norsemen. Many of them were killed in this manner for large rocks and large beams were thrown down upon them; great numbers also [were killed] by darts and spears and by every other means for killing man.

But the other forces, the Norsemen, were under the hurdles piercing the walls. What the Saxons and the Irishmen who were among them did was to throw large rocks so that they destroyed the hurdles over them. What they did in the face of this was to place large posts under the hurdles. What the Saxons did was to put all the ale and water of the town in the cauldrons of the town, to boil them and pour them over those who were under the hurdles so that the skins were stripped from them. The answer which the Norsemen gave to this was to spread hides on the hurdles. What the Saxons did was to let loose on the attacking force all the beehives in the town, so that they could not move their legs or hands from the great numbers of bees stinging them. Afterwards they left the city and abandoned it. It was not long after that [before they came] to wage battle again.

The story or *saga* of Ingimund thus not only tells of the peaceful Viking settlement of Wirral but also of its subsequent restlessness and of the start of attempts to acquire by force the city of Chester. The story ends on an ominous note:

> *It was not long after that before they came to wage battle again.*

The transcripts describing the story of Ingimund have themselves had a very interesting history. The particular part of the Irish annals containing the Ingimund story became known as the *Three Fragments*, which became lost. A vellum manuscript of unknown antiquity came into the possession of a certain Dubhaltach or Duald MacFirbis who made a copy. This then came into the possession of a Nehemias MacEgan.

Although that particular copy was also subsequently lost, another copy eventually found its way to Dublin scholar John O'Donovan who edited and published the story with the Irish Archaeological and Celtic Society in 1860. For the last 150 years the story which has come from "a copy of a copy of a vellum manuscript of unknown antiquity" has been under the scrutiny of scholars worldwide, but the general consensus now is that, despite the fictional nature of some of the text, the essence of the Ingimund story must be true. This conclusion is largely thanks to a penetrating and exhaustive study by Frederick Threlfall Wainwright in two papers: *North West Mercia*, and *Ingimund's Invasion*, both published in the 1940s, and reprinted in a collection of Wainwright's papers in 1975 and also in 2000 in *Wirral and its Viking Heritage*[18].

Ingimund's Story as recorded in these Irish sources is also supported by Welsh Chronicles which record the aborted attempt to settle in Anglesey. *Annales Cambriae* (entry for year AD 902)[19]:

> *Igmund came to Mona (Anglesey) and took Maes Osfeilion (Osmeliaun).*

Brut y Tywysogion (entry for AD 900)[20]:

> *900 was the year of Christ when Igmund came to the island of Anglesey and he held Maes Rhosmeilon.*

Although there is no direct record of him by the Anglo-Saxon chroniclers – who were more preoccupied by the Danes in the east - they do record the refortification of Chester by Æthelflæd in AD 907 (Figure 2-5) which is the same year as the attacks by Ingimund's forces as recorded in the Three Fragments. This was followed by construction of further forts in the

18. Cavill, P., Harding, S.E. and Jesch, J. (2000) *Wirral and its Viking Heritage*. English Place Name Society, Nottingham, UK, Chapters 3 and 4.
19. See Morris, J.D. translator (1980) *Nennius. The British History and the Welsh Annals*. Phillimore, Chichester, UK.
20. Jones, T. editor (1952) *Brut y Tywysogion, or the Chronicle of the Princes*. University of Wales, Cardiff.

surrounding region, including Eddisbury (AD 914), and Runcorn (AD 915) thereby containing this growing and restless colony from spreading deep into Cheshire: its growth, together with the steady build up of Scandinavian power in the north country resulting in the establishment of the Norse kingdom of York sparked a need for containment. J. M. Dodgson says the following[22]:

> *The urgency of the need for fortification on this frontier cannot have been lessened by the existence upon the frontier itself of restless Norse colonies, whose territories would serve as excellent beach-heads for any expedition striking down into Mercia along a short, direct and strategic route from Mersey.*

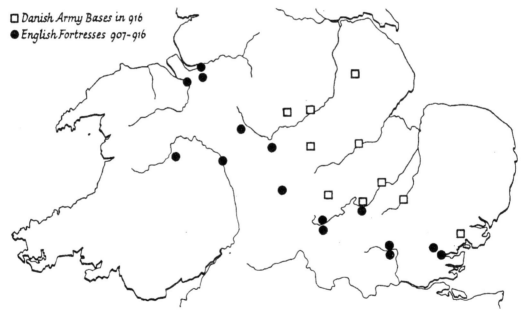

FIGURE 2-5. *Containing the menace of the Wirral Norsemen[21]. The bulk of the English defences were designed to contain the Danes to the east and north. Three forts constructed at Chester (907), Eddisbury (914) and Runcorn (915) were built for different purposes: to contain the Wirral Norsemen.*

21. From Wainwright, F.T. (1975) Æthelflaed, Lady of the Mercians. *In The Anglo-Saxons: Studies in Some Aspects of their History and Culture presented to Bruce Dickens.* Ed. P. Clemoes, pages 53-70, Bowes and Bowes, London.
22. Dodgson, J. M. (1957) The background of Brunanburh, *Saga-Book of the Viking Society* volume 14, pages 303-316.

Ragnald

If Ingimund is the great hulking figure of Dark Age Wirral then the corresponding figure for West Lancashire is another 10th Century Norse Viking who was operating in the Irish Sea region – Ragnald. Across the Mersey, West Lancashire was part of the Kingdom of Northumbria with its centre

at York: this was switching hands between the Danes and the English and then in AD 919 the historical records report that the Norseman Ragnald seized control. The University of Manchester historian N.J. Higham[23] has suggested that, following the expulsions from Dublin in AD 902, this Irish-Sea Viking settled and operated in the River Ribble area in parallel to Ingimund's activities in Wirral. He stayed in Ribble until AD 913-914 when he successfully launched a campaign against the then hostile Isle of Man: his power base then switched to there. From the Isle of Man he seized control of York in AD 919 establishing Northumbria (including Lancashire, which was then "South West Northumbria") as a Norse Kingdom.

FIGURE 2-6. *Signpost at Irby, Wirral. All the names are Norse or Norse influenced. Irby itself is a Norse name meaning "settlement of the Irish". Courtesy of Tom Holden.*

23. Higham, N.J. (1992) Northumbria, Mercia and the Irish Sea Norse, 893-926. In Graham-Campbell, J. ed. *Viking Treasure from the North West*, Chapter 3, National Museums and Guides on Merseyside, Liverpool.

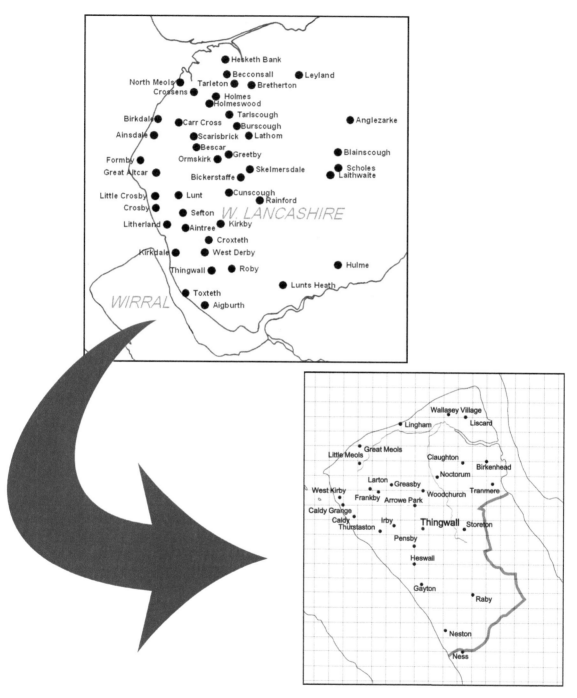

FIGURE 2-7. *Above: Names of towns or villages in West Lancashire and Wirral with Scandinavian roots. Facing: their putative original Scandinavian or "Hiberno-Norse" forms. Norse Þ is pronounced "Th". The meaning of the coastal region Þingsmere is explained on pages 31-32 as the "wetland or marshland associated with the Thing" warning travellers coming by boat to Wirral's Thing parliament, although its possible location indicated is speculative.*

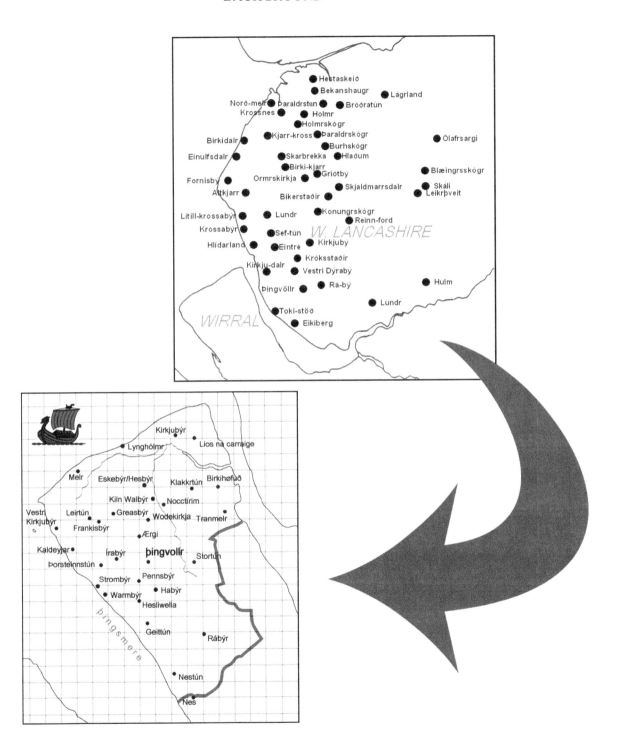

The place-name evidence

One of the strongest indicators of Viking settlement is place-names[24] (Figures 2-6, 2-7). There are many examples of places in Wirral and West Lancashire with the characteristic Scandinavian name element for a village or settlement – *by*. In Wirral the concentration of -by names is particulary dense - these include Frankby, Greasby, Irby, Raby, Pensby, Kirkby (the name for Wallasey Village until the 18th Century), West Kirby and Whitby, and just on the edge of Wirral there is Helsby. There are also some – *by* place-names that no longer exist – Kiln Walby (near what is now Upton), Haby (near Barnston), Eskeby or Askby (near Bidston), Stromby (near Thurstaston), Warmby (near Heswall) and Syllaby (near Saughall). In West Lancashire we have Crosby, Formby, Greetby (now part of Ormskirk), Kirkby, Roby and West Derby.

There are many other major names with Scandinavian or Hiberno-Norse (Irish-Norse) elements or origins. In Wirral we have Arrowe, Birkenhead, Claughton, Denhall, Gayton, Heswall, Hinderton, Larton, Lingham, Liscard, Meols, Mollington Torold (now Great Mollington), Ness, Neston, Noctorum, Shotwick, Storeton, Thingwall, Thurstaston, Tranmere and possibly Wodekirke (now Woodchurch).

In West Lancashire we have Aigburth, Ainsdale, Aintree, Altcar, Anglezarke, Argarmeols (now lost), Aynesargh (now lost), Becconsall, Bescar, Bickerstaffe, Birkdale, Blainscough, Breck House, Bretherton, Brettargh Holt, Burscough, Carr Cross, Crosby, Crossens, Croxteth, Cuerdley, Cunscough, Dalton and Dalton Lees, Drummersdale, Eggergarth, Eller Beck, Everton (possibly), Gunnolf's Moors, Harker, Harkirke, Haydock, Hesketh Bank (Moss and Sands), Holmes, Holmeswood, Hopecarr, Hoscar, Hulme, Kirkdale, Laithwaite, Lathom, Leyland, Limbrick, Litherland, Lunt, Lunt's Heath, Myckering, North Meols (and Meols Hall), Ormskirk, Rainford, Rainhill Stoops, Ravenmeols/ Raven Meal Hills, Ridgate, Roscoe Low, Sarscow, Scarisbrick, Scarth Hill, Scholes (3), Sefton, Skelmersdale, Snubscape, Sollom, Tarbock Green, Tarleton, Tarlscough, Thingwall, Toxteth, Ulnes Walton and Walton Breck/Warbreck.

24. Harding, S.E. (2000) *Ingimund's Saga*, Countyvise Ltd., Birkenhead, Chapter 3.

There are many English place-names of course, but for most of them there is little evidence to suggest they were in existence before the Vikings arrived, since the first records of their existence come later, in some cases much later.

Glossaries of these names – what they mean and where they are located – are provided in the books *Wirral and its Viking Heritage*[25] and *Viking Mersey*[26]. However, it is worth remarking on some here because of their special importance.

Thingwall

Thingwall (Figures 2-8 – 2-9) derives from Old Norse *þing-vollr* and means "Assembly Field" – the place of the Norse meeting place or parliament. The same name appears frequently in Norway, Denmark, Sweden and Iceland, such as Tingvoll near Molde in Norway and Þingvellir near Gull Foss (Figure 2-10) and near Stikkishólmur (Figure 2-11) in Iceland, and the *thing* element appears in the Althing – the present Icelandic parliament in Iceland, and also the Storting – the present parliament in Norway. Those in the British Isles have been reviewed by Professor Gillian Fellows-Jensen[27]. There is for example Tynwald in the Isle of Man, Tinwald near Dumfries in south western Scotland and Dingwall in Ross-shire, all known to have derived from these ancient Thingwalls or meeting places. The others are Tingwall in the Shetlands, Tingwall in the Orkneys, Tiongal in the Isle of Lewis, Dingwall in north-east Scotland, Tinwald in the Solway area of the Scottish borders and Tynwald in the Isle of Man – which still meets every 5th July. Of more dubious origins are Thingwala in Whitby, North Yorkshire and Dingbell Hill in Northumbria[28], but in England the only definite examples are Thingwall in Wirral and Thingwall Hall in South West Lancashire.

25. Cavill, P., Harding, S.E. and Jesch, J. (2000) *Wirral and its Viking Heritage*. English Place-Name Society, Nottingham, UK, Chapter 11.
26. Harding, S.E. (2002) *Viking Mersey*, Countyvise Ltd., Birkenhead, Chapters 4 & 7.
27. Fellows-Jensen, G. (1996) Tingwall: The significance of the name. In Waugh, D.J. and Smith, P. editors, *Shetlands Northern Links. Language and History*, Scottish Society for Northern Studies.
28. Only possibly.

FIGURE 2-8. *Thingwall in Wirral. Still a thriving village, the site of the Wirral Thing is believed to be just off the A551 at Cross Hill, with its brekka or slope. The arrow on the bottom photograph marks the possible point where the Lawspeaker would have addressed the Thingmen, or alternatively where the Thingmen would have sat as the Lawspeaker addressed them from below.*

FIGURE 2-9. *Thingwall in South West Lancashire, with Thingwall Hall now built on what may have been the Thing-brekka.*

FIGURE 2-10. *Thingvellir near Gullfoss in Iceland. There are several Thingvellirs in Iceland but this is the place of the original Althing[29].*

29. Courtesy of Ragnar Th. Sigurðsson.

FIGURE 2-11. *Site of another Thingvellir in North West Iceland near Stikkishólmur*[30] *as demonstrated by Saga enthusiast Birgir Jonsson of the University of Reykjavik.*

30. Courtesy of Birgir Jonsson, University of Reykjavik.

Raby and Roby

Raby in Wirral (Figure 2-12) and Roby in South West Lancashire both originate from the Old Norse rá-býr meaning boundary settlement and both appear to mark the boundaries of the (original) Norse population in those areas.

FIGURE 2-12. *Raby Mere in Wirral on the edge of the former settlement at Raby.*

Meols, North Meols and Tranmere

Meols in Wirral (Figure 2-13) and North Meols near Southport in West Lancashire both derive from the Old Norse *melr* meaning sandbank or sandhill. In the case of Meols in Wirral, up until the 18th Century there was a significant sandbank in the estuary adjacent to the old settlement forming a natural harbour. Archaeological evidence suggests the Romans took advantage of this and later when the Vikings arrived it became a major port or beach-market[31]. Dredging of the Mersey and coastal erosion led to a steady loss of this natural embankment and Meols lost its status as a port. Very low tides in the 19th Century revealed thousands of other archaeological finds, many dating from the Viking period as discussed below.

Tranmere on the Mersey estuary is a related place name and comes from Old Norse tran – meaning a cranebird or heron, so we have Tran-melr "sandbank with the cranebirds".

FIGURE 2-13. *Part of a 1732 map of Wirral showing the sandbanks at Meols forming a natural harbour*[32]. *The 18th Century forms of Tranmere, Thingwall, Wallasey Village (Kirkby) and Bromborough (Brunburgh) can be seen.*

31. Griffiths, D., Philpott, R.A. & Egan, G. (2007) *Meols, the Archaeology of the North Wirral Coast*, Oxford University School of Archaeology.
32. Courtesy of Dr. Paul Cavill, English Place Name Society, University of Nottingham.

Lathom and Litherland

Lathom (*hlaðum*, dative plural of ON *hlaða*) means "at the barns" and Litherland, preserves a pure Scandinavian genitive in *–ar* (*hlíðar*, genitive singular of ON *hlíð*), so Litherland – *hlíðarland* means "slope's land" or "land on a slope". These major settlement names, both in West Lancashire, are particularly noteworthy as they are symptomatic of a full Scandinavian language being spoken complete with inflections.

Irby, Liscard, Noctorum, Arrowe, Anglezarke, Brettargh

Place-names also indicate that many of the Norsemen came to Wirral and West Lancashire from Ireland, and brought Irish people with them. Prominent Irish names include Liscard (Old Irish *lios na carraige*, hall at the rock) and Noctorum (*cnocc-tírim*, hill that's dry). Irby is from the Old Norse Ira-byr, meaning settlement of the Irish or settlement of Norsemen coming from Ireland. Arrowe in Wirral and Anglezarke and Brettargh Holt in West Lancashire preserve the Hiberno-Norse *erg* or *ærgi* meaning "shieling", or "at the shieling" – meaning pastureland away from the farmhouse and derives from a type of farming called transhumance in which pastureland near the farmhouse is preserved from grazing for use as winter fodder.

The Irish influence also helped with the suggested explanation published in 2004 for the name Dingesmere from the Anglo Saxon Chronicle entry for AD 937 *The Battle of Brunanburh* as the "Thing's mere" – the wetland or marshland (Old Norse *marr*) associated with the Thing[33].

The Battle of Brunanburh

Brunanburh has considerable local, national and international significance. The battle took place in AD 937 between a new wave of Norse invaders coming from Dublin led by Olaf Guthfrithson joining forces with allies from Scotland led by Constantine against a Saxon-Mercian army coming from the

---

33. Cavill, P. Harding, S.E. and Jesch, J. (2004). Revisiting Dingesmere, *Journal of the English Place-Name Society*, volume 36, pages 25-38.

South led by Athelstan. The location of the Battle of Brunanburh - one of the most important battles in the history of the British Isles - has been the subject of stimulating and lively debate for over 100 years with nearly 40 different locations suggested. Two however stand out amongst the others. A compelling case has been made by Historian and Broadcaster Michael Wood for a location of the battle in the region of the Humber, on the basis of an account by the 12th Century historian John of Worcester that the invading forces came in via the Humber[34]. Another location currently favoured by many academics is for a location of the battle on Wirral. Brunanburh – the location of the battle as referred to in the contemporary Anglo-Saxon poem about the battle – happens to be the old name for Bromborough, and Dingesmere – the point of escape also referred to in the poem – has been explained as the "Things – mere or – marr", the wetland or marshland associated with the Thing – the Viking parliament, at Thingwall on Wirral[35] following a suggestion made by one of us (SEH) to Dr. Paul Cavill of the English Place Name Society in 2004. Thingsmere would have been a term used around that time by the local Norse and travellers coming by sea to visit the Thing. Nonetheless, until there has been a significant archaeological find we probably won't know for sure!

Minor place-names – evidence of a linguistically Scandinavian population

Whereas *major place-names* provide evidence of settlement – bearing Scandinavian elements which may have originated from the time of the settlement period, recent studies by scholars such as Kenneth Cameron[36] have shown that the *minor names* in an area tell us a great deal more about the vocabulary of the community. For example, in the case of Wirral, the intensity

34. Wood, M. (1980) Brunanburh revisited. *The Saga Book of the Viking Society*, volume 20, pages 200-217; and (2000) *In Search of England; Journeys into the English Past*, Penguin, London.

35. Cavill, P. Harding, S.E. and Jesch, J. (2004) Revisiting Dingesmere, *Journal of the English Place Name Society*, vol 36, pages 25-38.

36. Cameron, K. (1997) The Danish element in the minor and field-names of Yarborough Wapentake, Lincolnshire, in A.R. Rumble and A. D. Mills, eds, *Names, Places and People: An Onomastic Miscellany in Memory of John McNeal Dodgson*, Stamford, Paul Watkins, pages 19-25.

and distribution of minor place-name elements attests to the persistence of a Scandinavian-influenced dialect through the centuries, a persistence which may reflect the intensity of the original settlement[37]. Minor names such as those of fields, tracks, rivers or streams are less likely to have gone right back, as farmers/ fishermen in possession of a farm, field or coastal part may have altered the name as to their preference. However the existence of a name would have indicated the type of dialect word being spoken at the time that name was assigned. There are probably well over a thousand known examples in the Wirral and West Lancashire region. The interested reader is referred to the popular texts *Ingimund's Saga* or *Viking Mersey* where many of these names are given, together with their locations (or where these are not known precisely, estimated locations). For example, Wirral has nearly 100 examples of the element "rake" (Old Norse *rak*) which became a dialect word for a lane or track, over 50 examples of "carr" (Old Norse *kjarr*) meaning marshland or boggy land overgrown with brushwood, and 24 examples of "holm" (Old Norse *holmr*, island or dry ground in a marshy area) (Figure 2-14). By contrast, there is almost a complete absence of the corresponding English names – elements such as *mersc* (marsh) and *ēg* (dry ground in a marsh) – for the same features. The wealth of such names led F.T. Wainwright to make the following statements[38]:

"It is abundantly clear that we are dealing with an alien population of mass-migration proportions and not with a few military conquerors who usurped the choicest sites" and "They are incontrovertible proof that Scandinavians settled in great numbers".

37. Wainwright, F.T. (1943) Wirral Field Names, *Antiquity*, volume 27, pages 57-66; Harding, S.E. (2007) The Wirral carrs and holms, *Journal of the English Place Name Society*, volume 39, pages 45-57.
38. Wainwright, F.T. (1975) *Scandinavian England: Collected Papers*. Ed. Finberg H.P.R., Phillimore, Chichester, UK.

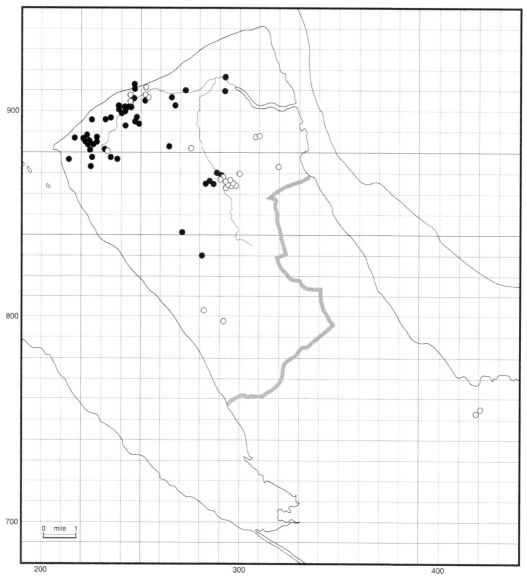

FIGURE 2-14. *Distribution of carrs (black circles) and holms (white circles) on Wirral, deriving from ON kjarr and holmr. There is a complete absence of the corresponding Old English forms.*

Linguistic evidence

Besides the minor place-names which were still in common use well into the 19th Century, the other evidence of a persistence of many words of Scandinavian origin centuries after the settlement period comes from some of the literature from the northwest of England, the most notable piece being the 14th Century poem *Sir Gawain and the Green Knight*. Although the poet is unknown, experts have identified he/she may have come from an area not far from Chester, and some have even linked the poem to Knight of the Garter Sir John Stanley (1345- 1413) of Storeton Hall, Wirral, who served in the Courts of both Richard II and Henry III. Part of the action described in the poem takes place in Wirral, and the incorporation of a large number of Norse dialect words makes the language of the poem distinct from Chaucer's Canterbury Tales – written around the same time. The Norse scholar Professor Nils-Lennart Johannesson, University of Stockholm[39], has identified the following list from the poem:

Words of Norse origin in Sir Gawain and the Green Knight

astryt: at once, straight away [OE *alswä* + ON *títt* 'often']
ay: always, ever [ON *ei*]
blande: mixture, together [OE *bland*, ON *í blana*]
bole: tree-trunk [ON *bolr*]
bonk: bank, hill, slope [ON *banki*, OIcel *bakki*]
boþe: both [OE *bä*, ON *báþi*]
boun: ready; dressed [ON *búinn*]
busk: get ready; dress [ON *búask*]
cayreõ: rides [ON *keyra*]
costes: condition, plight [OE *cost*, ON *kostr*]
cros: cross [ON *kross*]
derf: stout [ON *djarfr*]
dreped: killed [ON *drepa*]
dryõe: enduring [ON *drjúgr*]
felle: mountain [ON *fjall, fell*]
fro: from [ON *frá*]

---

39.   http://www.hf.ntnu.no/engelsk/staff/johannesson/!oe/texts/imed/07imed/07gls. htm

garysoun: treasure [OF *garisoun*, sense inflection by ON *gersumi*]

gate: road [ON *gata*]

gayn: advantage, a good thing [ON *gegn*]

gaynly: fitly, rightly [ON *gegn*]

gef: gave [ON *gefa, gaf*]

glent: glance [ON *glenta* 'to glance']

õeõe (after): cry (for) [ON *geyja*]

õette: grant [ON *játta*]

hendelayk: courtliness [ON *-leikr*]

karp: talk [ON *karpa* 'boast']

kest: to cast [ON *kasta*]

laykeõ: plays, amuses himself [ON *leika*]

lemed: shone [ON *ljóma*]

lygeõ: to lie (lay) [ON *liggja*]

meekly: meekly [ON *mjúkliga*]

menskful: gracious; term of address lady [ON *mennskr* 'human']

menskly: graciously [ON *mennskr*]

myre: mire, swamp [ON *mýrr*]

raged: ragged [ON *rôggvaðr*]

rapes: to hasten, hurry [ON *hrapa*]

renk: man, knight [ON *renk, rekkr*]

renneõ: runs [ON *renna*]

same, samen: together [ON *saman*]

semly: seemly [ON *sîmligr*]

sere: separate; several [ON *sér*]

skere: pure [ON *skærr*]

skyl: reason [ON *skil*]

sleõly: warily [ON *slægr*]

stor: strong, severe [ON *stórr*]

tok: took, **tan**: taken [ON *taka, tók, tekinn*]

tore: hard, difficult [ON *tórr*]

trayst: certain, sure [ON *treistr, treista* 'trusted', 'trust']

tyl: until [ON *til*]

þay: they [ON *þeir*]

vmbe: to be surrounded [OE *ymbe*, ON *umb*]

vn-slyõe: unwary [OE *un-* + ON *slægr*]

wale: to choose, waled: chosen [ON *velja, valdi*]

warþe: ford [OE *waroþ*: 'shore',with inflection by ON *vað* 'ford']
welcum: welcome [ON *velkominn*; OE *welcuma*]
won, wone: dwelling [ON *ván* 'hope']

ON: Old Norse; OE: Old English; OIcel: Old Icelandic

Archaeological evidence – Wirral

Norse Wirral is blessed with many fine examples of Viking Age stonework and treasure. This includes the magnificent hogback grave marker at St. Bridget's Church, West Kirby (recently beautifully restored by the Merseyside Conservation Centre)[40] (Figure 2-15) and another, smaller hogback from Bidston (Figure 1-5)[41]. Their dates of origin have been estimated as the early 11th century and the 10th century respectively. The Bidston find was very recent: local man Peter Crawford had discovered it in his garden and amateur archaeologist Jenny Whalley identified it as possibly Viking. She alerted an expert in Medieval stonework, Professor Richard Bailey from the University of Newcastle, who confirmed it as a Viking hogback, which according to a local newspaper report was one of the most exciting finds for many years[42].

Other fine examples include a large number of Viking artefacts discovered in low tides off the old Viking port of Meols. Although the majority of the finds were made in the 19th century they have only recently been catalogued[43] and include coins, Hiberno-Norse pins (Figure 1-4(a)), brooches, a drinking horn and what appear to be weapons from a possible pagan burial (Figure 2-16). Evidence was presented at a public meeting in 2008 of the remains of an elliptically shaped Viking house at Irby[44].

40. Collingwood, W.G. (1928). Early monuments of West Kirby, in John Brownbill ed. *West Kirby and Hilbre. A Parochial History*. Reprinted in Cavill, P., Harding, S.E and Jesch, J. (2000) *Wirral and its Viking Heritage*, English Place Name Society, Nottingham, UK, pages 84-97.
41. Bailey, R., Whalley, J., Bowden, A. and Tresise, G. (2006) A miniature Viking-Age hogback from the Wirral. *Antiquaries Journal*, volume 86, pages 345-356.
42. Powney, L. (2005) Stone most vital find in 20 years, *Wirral News Group*, November 30th.
43. Griffiths, D., Philpott, R.A. and Egan, G. (2007) *Meols, the Archaeology of the North Wirral Coast*, Oxford University School of Archaeology, pages 58-76.
44. Philpott, R.A. and Adams, M. (2010) *Irby, Wirral: Excavations on a Late Prehistoric, Romano-British and Medieval Site, 1987-1996*, National Museums Liverpool, UK.

Amongst other impressive evidence for the Viking settlements are remains of Viking crosses at West Kirby and Hilbre, at Woodchurch and at St. Barnabas Church in Bromborough[45] (Figure 2-17). At the Church of St. Mary and St. Helen at Neston there are seven fragments belonging to at least three Hiberno-Norse crosses, with fascinating imagery including the touching scene of a Viking couple embracing[46] (Figure 2-18). Three years after the Battle of Brunanburh, Olaf Guthfrithsson was king of Northumbria and York and intriguingly a coin (a round halfpenny) attributed to him was found in Neston[47] (Figure 2-19) shortly after a discovery of a silver ingot not far away[48] (Figure 2-20).

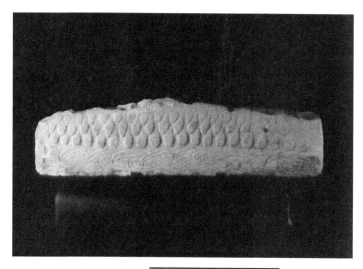

FIGURE 2-15.
Hogback tombstone at West Kirby[49].

45. Bu'Lock, J.D. (1958) Pre-Norman crosses of West Cheshire and the Norse settlements around the Irish Sea. *Transactions of the Lancashire and Cheshire Antiquarian Society* volume 68, pages 1-11. Reprinted in Cavill, P., Harding, S.E. and Jesch, J. (2000) *Wirral and its Viking Heritage*, English Place Name Society, Nottingham, UK, pages 70-83.

46. White, R.H. (1985) Norse period crosses at Neston, *Liverpool University Archaeological Newsletter* volume 1, pages 14-15 and (1985) *Liverpool University Archaeological Newsletter* volume 3, pages 1-2, and also (1986) Viking period sculpture at Neston, Chester, *Journal of the Chester Archaeological Society*, volume 69, pages 45-58.

47. Blackburn, M.A.S. (2006) Currency under the Vikings. Part 2: The two Scandinavian kingdoms of the Danelaw, c895-964, British Numismatic Journal, volume 76, pages 204-226.

48. Bean, S. (2000) Silver ingot from Ness, Wirral, in Cavill, P., Harding, S.E. and Jesch, J. *Wirral and its Viking Heritage*, English Place Name Society, Nottingham, UK, pages 17-18.

49. Courtesy of the National Conservation Centre, Liverpool, who have beautifully restored this piece.

FIGURE 2-16. *Meols: Weaponry from a Viking warrior. Top: bent spear head. Middle: shield boss. Bottom: axe head. Courtesy of Dr. David Griffiths, University of Oxford.*

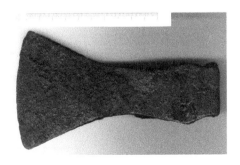

FIGURE 2-17.
*Hiberno-Norse
Ring headed cross,
Bromborough*[50].

50. Restored ring-headed church cross, Bromborough. This cross was re-erected at St. Barnabas's church on St. Barnabas's Day, 11th June 1958. See Bu'Lock, J.D. (1958) Pre-Norman crosses of West Cheshire and the Norse settlements around the Irish Sea. *Transactions of the Lancashire and Cheshire Antiquarian Society* volume 68, pages 1-11. Reproduced in Cavill, P., Harding, S. and Jesch, J. (2000) *Wirral and its Viking Heritage*, English Place Name Society, Nottingham, UK, pages 70-83.

FIGURE 2-18. *Hiberno-Norse cross fragment at Neston showing men fighting on horseback on one side (a) and part of a couple embracing on the other (b), seen better with laser reconstruction by the National Conservation Centre at Liverpool (c). This is just one of seven fragments found which appear to have belonged to at least three crosses*[51].

51. Courtesy of Dr. Roger White, University of Birmingham.

FIGURE 2-19. *King Olaf Guthfrithsson coin (AD 941), Neston*[52].

---

52. Courtesy of the Fitzwilliam Museum, Cambridge.

FIGURE 2-20. *Silver ingot, Ness*[53].

53. Courtesy of Dr. Simon Bean, National Museums of Liverpool.

Chester

Just to the south of Wirral at Castle Esplanade in Chester a hoard of 520 coins was discovered in 1950 (marked with the names of Scandinavian and English moneyers) together with a quantity of hack-silver and a number of fragmentary or whole silver ingots of the Ness type (Figure 2-21 and 2-22), and other artefacts have been found such as a ring pin and brooch (Fig 2-23). The hoard has been dated at around AD 965. Although Ingimund's Story (see pages 16-18) stops short of the Wirral Vikings gaining a foothold in Chester it is clear from the archaeology and the street names that their presence became significant particularly in the southern part of the city. In the 10th and 11th Centuries Chester developed into a financial centre with its own mint, and approximately 25% of the City moneyers from that time bore Scandinavian names. A proportion of the 10th-century moneyers from the area bore clear Irish-Norse origins with names such as *Irfara* (ON derivation meaning "Ireland journeyer"), *Oslac* and *Mældomen*. Off what is now Lower Bridge Street – not far from Castle Esplanade - stands St. Olaves church, probably located on the same site as an original church constructed by the 11th Century Scandinavian community.

FIGURE 2-21. *Part of the Castle Esplanade Hoard, Chester, consisting of silver pennies, ingots and hack-silver – pieces of jewellery chopped up to act as currency. Everything was contained in the small cooking pot which stands in the background. The hoard had either been lost or deliberately buried by a Viking trader some time around AD 965.*[54]

54. Courtesy of Chester City Council/Grosvenor Museum.

FIGURE 2-22. *Closer view of the Castle Esplanade Hoard.*

(a)

FIGURE 2-23. *Viking ring pin (a) and brooch (b) also found in Chester.*

(b)

On the outside of Chester lies another outstanding archaeological find. In 2004 metal detector enthusiasts in the village of Huxley discovered buried in a field another huge hoard of Viking treasure – 21 flattened silver bracelets, a silver ingot, and 39 lead pieces, identified a short time afterwards by Chester Archaeologist Dan Garner. This find was also dated to around the first years of the 10th Century, not long after Ingimund's arrival in Wirral in AD 902. The "Huxley hoard" (Figure 2-24) has been on display in London, Liverpool and Chester and details of the find have recently been published by the National Museums of Liverpool[55].

FIGURE 2-24. *Part of the Huxley Hoard of Viking Treasure, Chester. The hoard was buried around AD 905 and discovered by metal detector enthusiasts at a rally in 2004. The bracelets at the top are approximately 11.5 cm long. Courtesy of David Flower © National Museums Liverpool (Museum of Liverpool).*

---

55. Graham Campbell J.A. and Philpott, R.A. eds. (2010) *The Huxley Viking Hoard. Scandinavian Settlement in the North West*, National Museums of Liverpool.

Archaeological evidence – West Lancashire

West Lancashire boasts one of the most important discoveries of a Viking treasure hoard outside Scandinavia, a find made in 1840 by workers on the banks of the River Ribble at Cuerdale. *The Cuerdale Hoard* (Figures 1-4b and 2-25) has been dated to about AD 905, around the same time as the Ingimund settlements in Wirral, and about the same time as the deposition of the Huxley hoard. Much more recently a lead weight with typical Viking interlace design was discovered not far away in Preston (Figure 2-26).

FIGURE 2-25. *Selection from the Cuerdale Hoard. Approximately 8,600 bullion pieces and coins were found in 1840 – the largest known silver hoard in the Viking world. Buried ~ AD 905 - about the same time as the Huxley Hoard - and now displayed in the Grosvenor Museum in Chester and the British Museum*[56].

---

56. Courtesy of the British Museum.

FIGURE 2-26. *Three different views of a lead weight with interlace found near Preston*[57].

Another important find – a Viking coin hoard with up to 300 pieces - was made much earlier in 1611 at Harkirke near Crosby[58] (Figure 2-27). Although the coins are now lost a copperplate of 35 of the coins was made by one of the discoverers and has been on display at the British Library in London.

Although West Lancashire does not have the same impressive array of Viking-age stonework as that found in Wirral, an intriguing discovery was made at Prescot on the outskirts of Liverpool – what appeared to be pieces of the Viking board game Hnefatafl or "check table" now beautifully restored and displayed in Warrington Museum[59] (Figure 2-28).

57. Courtesy of Dr. David Griffiths, University of Oxford.
58. See Edwards, B.J.N. (1998) *Vikings in North West England: the Artifacts*. Centre for North West Regional Studies, University of Lancaster.
59. Bu'Lock, J.D. 1972 (*Pre-Conquest Cheshire 383-1066*). Cheshire Community Council, Chester, and Harding, S.E. (2000) *Ingimund's Saga: Norwegian Wirral*. Countyvise Ltd., Birkenhead.

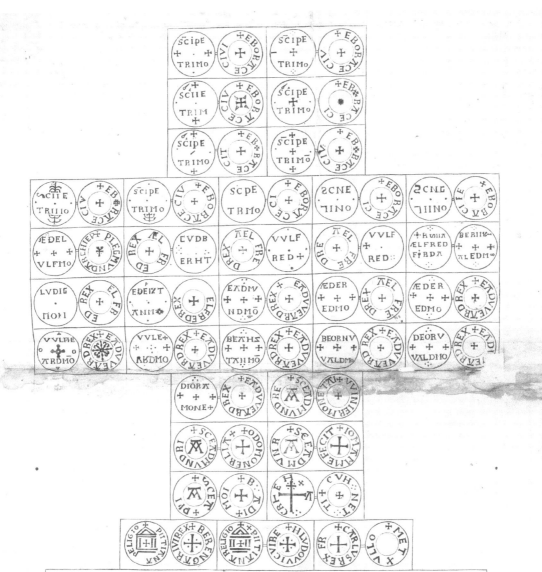

FIGURE 2-27. *Copperplate print depicting coins from Harkirke, West Lancashire, found in 1611 (British Library, Harleian MS 1437, 128v-129r)[60]. The engraved plate was supposedly made in 1613, and is now in the Lancashire Record Office.*

60. Courtesy of the British Library.

Describing the main hnefi piece, the museum writes the following: "Jet figure with bevelled edges, ornamented with scratched lines and circles, possibly drawn with an instrument such as a compass". Of the other piece "Jet figure: cylindrical with bevelled top".

FIGURE 2-28. *Hnefatafl or "Tablut" pieces from about the 10th-Century now at Warrington Museum*[61]. *The drawing pin is included to give an idea of size.*

---

61. Courtesy of Angela Doyle and Keith Scott, Warrington Museum.

Chapter 3
DNA: MATERNAL AND PATERNAL ANCESTRY

The remarkable molecule DNA – deoxyribonucleic acid – contains the genetic code for what we are. It provides the message or template from which proteins in cells are made, which in turn determine many of our characteristics such as our sex, height, skin, hair and eye colouring, and the risk of developing some diseases. Because it passes down from generation to generation, it also provides important clues as to where our ancestors came from.

The structure of DNA was solved in 1953 by James D. Watson – a junior Research Fellow – and Francis H. Crick – a PhD student – at the Cavendish Laboratory in the University of Cambridge. The DNA molecule consists of a series of chemical building blocks or "bases" carried by two backbone polymer chains running in opposite directions and twisted into a double-helical structure[62] (Figure 3-1). Pairs of bases in the centre of the molecule are attracted to each other by what we call hydrogen bonds which had been discovered earlier in DNA by Michael Creeth (Figure 3-2), a young PhD student supervised by J.M. Gulland and D.O. Jordan at University College Nottingham, and these bonds hold the molecule together[63]. There are four bases, adenine (A), cytosine (C), thymine (T) and guanine (G). Adenines on one chain pair with thymines on the other and guanines pair with cytosines to form the double-stranded molecule. It is the sequence or order of the base pairs which forms the genetic code or template of life – the sequence of A, C, T and G units.

62. Watson, J. D. and Crick F.H.C. (1953) A structure for deoxyribose nucleic acid, *Nature*, volume 171, pages 737-738.
63. Creeth, J.M., Gulland, J.M. and Jordan, D.O. (1947) Deoxypentose nucleic acids, *Journal of the Chemical Society*, article 214, pages 1141-1145 and Creeth, J.M. (1947) *Some Physico-chemical Studies on Nucleic Acids and Related Substances*, PhD Dissertation, University College Nottingham – a double stranded structure of DNA was for the first time presented here. This work was recently reviewed by Harding, S.E. and Winzor, D.J. (2010) James Michael Creeth, 1924-2010, *Macromolecular Bioscience*, volume 10, pages 696-699.

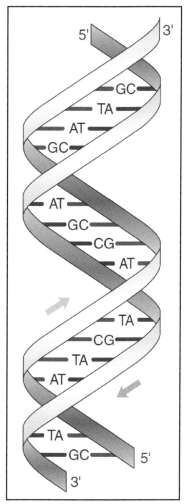

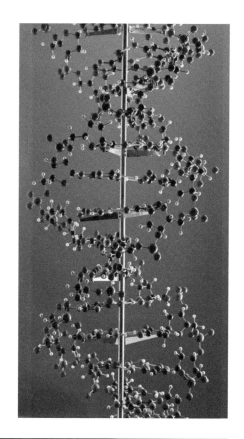

FIGURE 3-1. *Top Left: DNA, the molecule of life, consists of a double-helical structure with chemical building blocks or bases A, T, C and G holding the molecule together[64]. Top Right: Ball and spoke model of DNA built by Alexander Barker, for the International Science Pavilion, Brussels World Exhibition, 1958[65]. Bottom Right: Watson and Crick drinking coffee at the Cavendish Laboratory after their momentous discovery in 1953[66].*

64. Copyright 2004 from *Human Evolutionary Genetics* by Jobling, M.A., Hurles, M.E. and Tyler-Smith, C. Reproduced by permission of Taylor and Francis Group, LLC, a division of Informa plc.

65. Reproduced courtesy of the Laboratory of Molecular Biology, Cambridge.

66. Photo by Antony Barrington Brown and reproduced with permission from the Science Photo Library.

FIGURE 3-2. *PhD student J. Michael Creeth whose hydrodynamic experiments in 1947 indicated a two-strand structure for the DNA molecule held together by hydrogen bonds: a finding crucial to the discovery of the structure of DNA by Watson and Crick.*

The DNA molecules in humans are packaged into structures called chromosomes which exist in every human cell (apart from red blood cells). There are 46 of these chromosomes arranged into 23 pairs (Figure 3-3). In 22 of these, members of each pair look very similar to each other under a microscope and are known as autosomes. The remaining pair are what we call the sex chromosomes. In females they also look very similar to each other, and are called X-chromosomes, whereas in males they are different - one being an X-chromosome, and the other a much shorter chromosome known as a Y-chromosome. Essentially, if you have a Y-chromosome you are male, and if you do not you are female. There is also a small amount of DNA found outside the cell nucleus in structures called mitochondria – these are the parts of the human cell dedicated to the generation of energy for driving the processes of life.

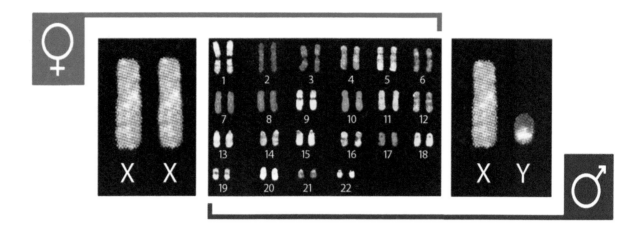

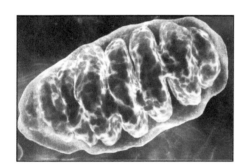

FIGURE 3-3. *Human chromosomes. Our DNA is packaged in the cell into 23 pairs of chromosomes. There is one pair that determines our sex – whether we are male or female. In females the two sex chromosomes are similar and called X chromosomes, while in males there is one X and a shorter chromosome called Y. A small amount of our DNA is not contained in chromosomes but in structures called mitochondria, as seen in the computer art representation on the left. A mitochondrion is typically 1/2000th of a mm in size. Reproduced with permission of Hybrid Medical Animation and the Science Photo Library.*

We receive approximately 50% of our DNA from our mothers and 50% from our fathers, with most of the DNA from each being shuffled or recombined between the chromosome pairs when eggs and sperm are made.

DNA - Messages from our ancestors

DNA – bequeathed to us from our ancestors - is a 'text' that has changed slowly through time, and varies among individuals. To study the DNA from our ancestors we could analyse DNA from skeletons. This provides 'real' information about the past. There is however a major drawback in that it is very difficult to extract useful amounts of DNA, sample sizes of individuals are usually small, the samples are prone to modern DNA contamination (which confuses the results) – and there may actually be no modern descendants that relate to the skeletal DNA.

The alternative approach is to analyse the DNA from modern people – here it is easy to get samples, provided that the necessary ethical approvals have been received. The drawback here is that the samples taken can be unrepresentative of past populations – we therefore need to choose which individuals to sample carefully, and to use statistical methods to extrapolate the genetic patterns seen in modern populations back into the past.

The Genetics of Physical Characteristics

What kinds of DNA variation can we study? As we will see below, most of the variation in DNA has no effect on our physical characteristics (it is termed 'neutral'). However, some DNA variation does have effects that we can measure or observe directly.

Blood groups – which are inherited and determined by genes - have for many years been considered as a marker of population ancestry[67] – but are generally poorly discriminating and widespread in many different populations.

Other physical characteristics are in people's general appearance – pigmentation, stature, and facial shape, for example. Such characteristics have also been studied for many years[68], starting in the 19th Century with, for example, the work of Charles Darwin's cousin Francis Galton (who also touched on the controversial characteristic of intelligence) and more recently in a survey of facial shape across Britain conducted by Channel 4 TV and reported by presenter Neil Oliver.

Although our understanding of pigmentation is improving rapidly, the genetic basis of most common physical characteristics such as facial shape remains obscure. In many cases – stature, for example - environment plays a major role, together with many poorly understood genes. There is a wide distribution of the characteristics, and their underlying genetic factors, across northern Europe. Nonetheless the stereotypical perception of a

67. Cavalli-Sforza, L.L., Menozzi, P. and Piazza, A. (1994). *The History and Geography of Human Genes*, Princeton University Press, Princeton, New Jersey.
68. Geipel, J. (1969) *The Europeans: an Ethnohistorical Survey*. Longmans, London.

"Scandinavian" as fair haired, oval-faced, blue-eyed and (the men, at least) tall has been based on a perceived higher proportion of these "phenotypes" in Scandinavia.

Fair hair and blue eyes

The highest proportion of people with fair/blond hair and blue eyes appears to be found in central Sweden, Norway and Finland, the Baltic states and the northern parts of Poland and the former German Democratic Republic with a reported 70-80% people possessing these phenotypes[69,70], (Figures 3-4): one can reasonably infer the origins for both are in the Sweden-Baltic sea region The genes responsible are now becoming increasingly better understood[71,72]. The *DeCode* organization[73], with its headquarters based in Reykjavik, Iceland, has been a world leader in studying the human genome based on the analysis of the DNA from thousands of individuals, and is now able to make predictions about a person's eye and hair colour on the basis of their DNA.

There has clearly been progress, and we have a better understanding of the genes responsible for pigmentation differences; however, at the moment it is not possible to correlate the distributions of the underlying genetic variants behind hair and eye colour to the Viking migrations.

69. Beals, R.L. and Hoijer, H. (1965) *An Introduction to Anthropology* (3rd edition), Macmillan, New York.

70. Frost, P. (2006) European hair and eye color. A case of frequency-dependent sexual selection? *Evolution and Human Behaviour*, volume 27, pages 85-103.

71. Sulem, P., Gudbjartsson, D.F., Stacey, S.N., Helgason, A., Rafnar, T., Magnusson, K.P., Manolescu, A., Karason, A., Palsson, A., Thorleifsson, G., Jakobsdottir, M., Steinberg, S., Pálsson, S., Jonasson, F., Sigurgeirsson, B., Thorisdottir, K., Ragnarsson, R., Benediktsdottir, K.R., Aben, K.K., Kiemeney, L.A, Olafsson, J.H., Gulcher, J., Kong, A., Thorsteinsdottir, U. and Stefansson, K. (2007) Genetic determinants of hair, eye and skin pigmentation in Europeans, *Nature Genetics* volume 39, pages 1443-1452.

72. Eiberg, H., Troelsen, J., Nielsen, M., Mikkelsen, A., Mengel-From, J., Kjaer, K.K. and Hansen, L. (2008) Blue eye color in humans may be caused by a perfectly associated founder mutation in a regulatory element located within the HERC2 gene inhibiting OCA2 expression, *Human Genetics*, volume 123, pages 177-187.

73. http://www.decode.com/research/

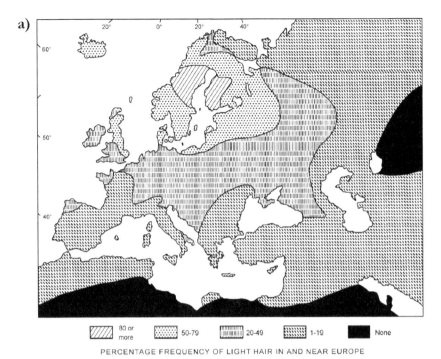

PERCENTAGE FREQUENCY OF LIGHT HAIR IN AND NEAR EUROPE

FIGURE 3-4.

Distribution of people with light or fair hair (a) and light or blue eyes (b) in Europe.[74]

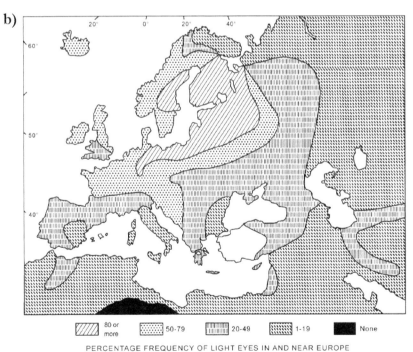

PERCENTAGE FREQUENCY OF LIGHT EYES IN AND NEAR EUROPE

74. From Beals, R.L. and Hoijer, H. (1965) *An Introduction to Anthropology*, 3rd edn. MacMillan, New York, USA, p214 and reprinted with permission of Pearson Education.

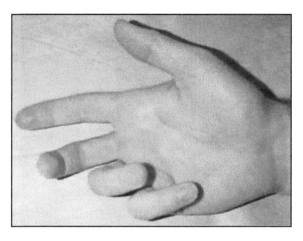

FIGURE 3-5. *Dupuytren's contracture.*
A tightening of the elastic tissue underneath the skin means that it becomes difficult, sometimes impossible, to extend one or more of the fingers of a hand (commonly the 4th and 5th fingers). Common in Scandinavia, and in those parts of Britain influenced by the Vikings[75].

Dupuytren's Contracture

Another physical characteristic that has been widely considered as a marker of Scandinavian ancestry is a physical condition of the hand known as Dupuytren's contracture (Figure 3-5). In this condition (three times more common in men than in women) a person is unable to extend without difficulty one or more fingers of a hand – normally the 4th & 5th fingers. This is due to a condition of the elastic tissue underlying the surface of the skin. Although 'sporadic' cases occur, many cases are inherited in a simple way suggesting a single causative gene acting in a "dominant" manner – that is to say that if a person receives the gene for the contracture from one parent and the gene for non-contracture from the other, he will have the contracture. In other words only one copy of the gene is needed to give someone Dupuytren's.

The relatively high proportion of people in Scandinavia having this condition suggests a possible Viking origin for a large proportion of Dupuytren's cases. There is even evidence of this condition amongst Vikings in the Icelandic sagas[76]. As with the fair-hair - blue eyes combination in Britain

75. From http://www.med.und.edu/users/jwhiting/duprr.html and reproduced courtesy of Dr. Jeff Whiting, Saint Louis University School of Medicine.
76. The *Longer Saga of Magnus of Orkney* tells about a man called Sigurdr who after a pilgrimage to the shrine of Holy Magnus allegedly had a complete recovery – the fingers became supple and flexible and could be put to any use. See Whaley, D.C. and Elliot, D. (1993) Dupuytren's disease: a legacy of the north? *Journal of Hand Surgery*, British and European volume 18, pages 363-367.

the incidence of Dupuytren's contracture is anecdotally more frequent in regions of Britain influenced by the Vikings, and in a very unscientific show of hands taken at the 2007 Knowsley Hall meetings announcing the genetic survey results a large number of people showed they had the contracture.

However, the genetic basis of the condition is currently unknown[77] and it also appears in other populations with no clear connections with Scandinavia. If an underlying genetic mutation responsible for the Scandinavian form is identified in the future, it could become an interesting genetic marker for Scandinavian ancestry in British populations.

DNA: maternal and paternal lines

Because of the difficulties of using genetic variation connected to physical differences among people, scientists commonly focus on the plentiful 'neutral' variations in DNA that have no discernible effect on the person who carries them, but which can provide evidence about their ancestors.

To use genetics to learn about ancestry we consider characteristic features or 'markers', namely the occurrence of particular bases at particular locations within our DNA, or particular patterns of short repeating units in DNA. However, the patterns of such variants in most of our DNA are difficult to interpret, as every generation it undergoes reshuffling or 'recombination' which blurs the signal of ancestry.

However, there are two parts of our DNA which we receive from our parents that are not reshuffled (Figure 3-3). Most of the DNA on the Y-chromosome is passed down from father to son essentially unchanged – except for rare mutations which can occur over the course of hundreds or thousands of years. Similarly, the DNA in the mitochondria is passed down the maternal line over very many generations with little or no change. So by characterising the type of DNA a man has in his Y-chromosome and in his mitochondria we can obtain specific information about his paternal and maternal ancestry. With women we can also enquire about ancestry but only

77. A recent study on one particular family has suggested that DNA on chromosome #16 is involved.

along the maternal line as women do not have Y-chromosomes.

It has to be stressed that this method of enquiry can only tell us about one or two lines of ancestry (Figure 3-6). If we go back 1000 years we have hundreds of thousands of ancestors as the diagram suggests, and we learn nothing about the great majority of these by testing Y-chromosomes and mitochondrial DNAs.

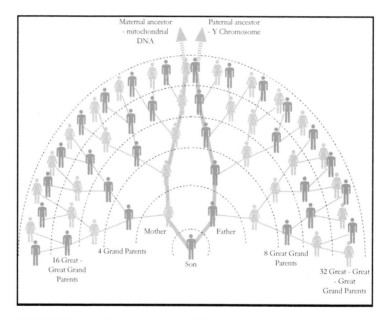

FIGURE 3-6. *Multiple ancestors. DNA testing of uniparentally inherited markers (Y and mtDNA) can only provide information about one line of ancestry for women and only two lines for men.*

Y-chromosome Jargon

Base: Of 4 types: adenine (symbol A), thymine (symbol T), cytosine (C), and guanine (G). These are linked together into a long chain. The sequence of bases on the chain characterizes a person's DNA.

Base-pair: DNA has two long chains of bases wrapped around each other into a double helix in such a way that the A residue of one chain pairs with a T on the other chain, and a C residue pairs with a G.

Binary marker: A difference between Y-chromosomes usually involving the substitution of one base for another (an SNP); the process of change at such markers is very slow. Groups of chromosomes characterized by these kinds of markers can be very plentiful in particular populations, and so they are not the markers of choice for discriminating between individuals.

Haplogroup: Overall classification of a man's Y-chromosome based on a set of binary markers.

Haplotype: Detailed classification of a man's Y-chromosome based on the numbers of tandem repeats for a chosen set of at least 6 STRs.

PCR: Polymerase Chain Reaction. Technique which amplifies sections of DNA from an individual, allowing polymorphisms such as STRs or binary markers to be analysed.

Polymorphism: A difference between the DNA of two Y-chromosomes due to a rare error in the genetic replication machinery.

SNP: Single nucleotide polymorphism. The commonest kind of binary marker, involving the substitution of one base by another, or sometimes the deletion or insertion of a single base.

STR: Short tandem repeat (sometimes called a microsatellite). A sequence of 3-5 bases, e.g. TAGA, which is repeated a variable number of times (typically 8-30) in a row on a man's Y-chromosome. This kind of marker changes relatively rapidly between the generations, and is useful for discriminating between individuals.

The male Y-chromosome: haplogroups

The potential for using the male Y-chromosome as a tool for genealogical research was first revealed with two research papers published in 1985[78], but it was not until the turn of the millennium that the technology had advanced significantly to enable the easy study of people's DNA as a means of throwing light on the origins and migrations of populations. Advances are being made all the time, making it an increasingly powerful probe into the past.

As indicated above the Y-chromosome is the reason that a man is male – it carries a gene which switches on the pathway of male sex determination early in development. The Y-chromosome (Figure 3-3) consists of a huge length of DNA (deoxyribonucleic acid) with about 60,000,000 base pairs of A-T or C-G on its two anti-parallel chains. In common with the other chromosomes the order in which the bases occur characterize a man's Y-chromosome. Apart from the essential sex-determining gene, very little of the DNA on the Y-chromosome defines anything about how we look or what diseases we might suffer from, and most is what the experts call "junk DNA". However, as we have indicated above, unlike DNA on the autosomes and X-chromosomes, the DNA on the Y-chromosome is passed from generation to generation only along the male line, from father to son. Another important difference between the Y and the other chromosomes is that most of the DNA on it escapes from a reshuffling process (called recombination) which occurs every generation. This means that the message in the DNA sequence that we get from our Y-chromosomal ancestors is much clearer to read than the message from the ancestors who bequeathed their autosomes and X-chromosomes to us. Changes on Y-chromosomes only occur through a mistake or "mutation" in the body's normally impeccable copying machinery, and this produces a difference between

---

78. Cassanova, M. Leroy, P., Boucekkine, C., Weissenbach, J.,. Bishop, C., Fellous, M., Purrello, M., Fiori, G. & Siniscalco,. M. (1985) A human Y-linked DNA polymorphism and its potential for estimating genetic and evolutionary distance *Science*, volume 230, pages 1403-1406 and Lucotte, G. and Ngo, N.Y (1985) P49f, a highly polymorphic probe that detects TaqI RFLPs on the human Y chromosome *Nucleic Acids Research*, volume 13, page 8285. For a review, see Jobling, M.A. and Tyler-Smith, C. (1995) Fathers and sons: the Y chromosome and human evolution. *Trends in Genetics*, volume 11, pages 449-456.

different Y-chromosomes in the population, known as a polymorphism. These changes are however very rare – some occur over a course of hundreds of years, and some over thousands or tens of thousands of years or longer. Some (such as base substitutions, where one letter of the DNA code changes for another) have probably happened only once in human history – these are sometimes called unique event polymorphisms or "single nucleotide polymorphisms" – SNPs for short (Figure 3-7 and 3-8). It is how many and where these SNPs are on a man's Y-DNA which determine his haplogroup. Once such a mutation or polymorphism has occurred, and the man whose Y-DNA has undergone this harmless mutation has a son or sons, then this change becomes preserved and indeed propagated.

Using statistical methods it is possible to gauge an approximate date when the first change that we now see preserved among modern men took place - about 90,000 years ago - and the person who bequeathed this change is known as the Y-chromosome most recent common ancestor or "MRCA" which some have called "Y-chromosomal Adam". Descending from this ancestor there are 20 main haplogroups corresponding to SNPs occurring at different places along the Y-DNA (Figure 3-7) – these are given the names A, B, C, D, E, F*, G, H, I, J, K*, L, M, N, O, P*, Q, R, S and T. These haplogroups comprise a 'tree' of all Y-chromosome types, in which the types descending most directly from Y-chromosomal Adam (the root of the tree) are almost exclusively found in Africa, providing support for the idea that modern humans originated there. The SNP mutations on the DNA which gave rise to these haplogroups are given codes such as M89, M173 etcetera. The "*" means a collection of haplogroups related by a particular mutation such as F* (mutation M89), K* (mutation M9), P* (mutation M45) and DE* (mutation M145).

All of these SNPs have arisen over a time period of thousands or tens of thousands of years. Population geneticists have been able to assess when these changes have taken place, by assessing the variation within each haplogroup using the more rapidly changing markers, STRs.

After a particular SNP had been established then at some stage later a further SNP could have occurred elsewhere on the Y-DNA. Again, so long as the male who first carried this mutation himself had sons and so on,

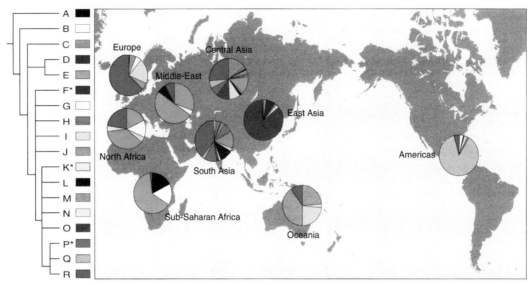

FIGURE 3-7. *The main Y-chromosome haplogroups and their distribution through the world. The earliest-branching is A and the branches to the other main haplogroups correspond to different single nucleotide polymorphisms (SNPs), spanning tens of thousands of years.*[79]

this minor haplogroup SNP (sometimes called a "subclade") would have become established in the male population.

For example, in haplogroup R, defined by the SNP M207, a further SNP (M173) established itself, and then two further ones occurred leading to R1a and R1b. Then more recently the Y-chromosome DNA of a man possessing R1b, for example, underwent a further mutation leading to further diversity to give the subgroups R1b1 etc. Similarly, within haplogroup I, further SNPs have led to subgroups called I1, I1a, I2b, I2b1, and so on. In the same way that geneticists group collections of major haplogroups together into superhaplogroups (represented by the * notation), they sometimes do this with the minor haplogroups or subclades such as R1* (linked by the mutation M173), which includes not only R1 but also R1a1 (commonly known as R1a) and R1b1b2, (commonly referred to as "R1b") etc. One other piece of jargon they use is the prefix "x" meaning "excluding", so that K*(xR1) means K* superhaplogroup excluding R1, and R1*(xR1a) means the R1* superhaplogroup excluding R1a.

---

79. Copyright 2004 from *Human Evolutionary Genetics* by Jobling, M.A., Hurles, M.E. and Tyler-Smith, C. Reproduced by permission of Taylor and Francis Group, LLC, a division of Informa plc.

One source of confusion about haplogroup names is that they change as new SNPs are discovered or new populations surveyed. For example, the mutation M269 defines a subclade of haplogroup R that was originally known as R1b3, but is currently called R1b1b2. Here, we use nomenclature agreed in 2008[80].

The commonest haplogroups found in the British Isles and Western Europe

Y haplogroups are very non-randomly distributed in different regions and populations. For example, in Europe (Figure 3-8), we find high frequencies of Y chromosomes within haplogroups R, I, E, and J, together with lower frequencies within the main haplogroups G, N, Q and T. However, it is extremely rare to find chromosomes within haplogroups such as A, B, C, D, L, M and O. This geographical specificity extends to more regional scales as we shall see, and this is why the distributions of Y types can tell us something about past migrations.

More specifically, the common haplogroups within Europe and the British Isles are as follows:

Haplogroup R1b1b2 is the commonest Y chromosome group in Western and Northern Europe, and is the most common group in Scandinavia. It follows a rising southeast to northwest gradient of frequency from Turkey to Ireland (>80%), and in our view represents the spread of farming populations that began 10,000 years ago, in the Neolithic period (New Stone Age).

Haplogroup I1. Part of the broader lineage haplogroup I, it represents the indigenous hunter-gatherer populations that were present in Europe before the farming populations that began 10,000 years ago, in the Neolithic period (New Stone Age). It is also a possible signal of Viking ancestry, since it is found at its highest frequencies in Scandinavia.

---

80. Based on Karafet, T.M, Mendez, F.L, Meilerman, M, Underhill, P.A., Zegura, S.L. and Hammer, M.F. (2008) New binary polymorphisms reshape and increase resolution of the human Y-chromosomal haplogroup tree. *Genome Research*, volume18, pages 830-838.

Haplogroups I1a, **I2b** and **I2b1** are rarer subgroups of haplogroup I, and found at lower frequencies in Britain, as well as Central and Southern Europe.

Haplogroup E1b1b1 is generally quite rare in Britain, being found at greatest frequency around the Mediterranean. It also represents part of the Neolithic spread of farming populations. A famous carrier of a haplogroup E1b1b1 Y chromosome is Sir David Attenborough.

Haplogroups J2a1b and **J2a1h** are further vestiges of the Neolithic expansion. They are generally rare in Britain, being found at greatest frequency around the Mediterranean.

Haplogroup R1a1 is generally quite rare in Britain except in regions with strong Norse ancestry. It is often regarded as a signature of Norse Viking ancestry, and is common in Norway. It is also found at high frequencies in Central Europe and India.

Haplogroup N is very rare in Britain, but found at high frequencies in northern Scandinavia and east of the Baltic. It was spread from north-east Asia with so-called 'Finno-Ugric' languages, and in Britain could be a signal of Viking ancestry.

Haplogroup G2a is also generally quite rare in Britain, and in Europe overall (5%), being most common in some Mediterranean regions. Like haplogroup I, it probably reflects the indigenous hunter-gatherer populations that were present in Europe before the Neolithic age.

Haplotypes – probing diversity within haplogroups

Although some rare haplogroups are quite sensitive in discriminating within populations, other haplogroups by themselves are very poorly discriminating, the most notable being R1b1b2 which although the most common haplogroup across much of Scandinavia, is also the commonest across most of Western Europe. Fortunately there is a more sensitive type of difference in Y-chromosomes arising not from a base substitution but

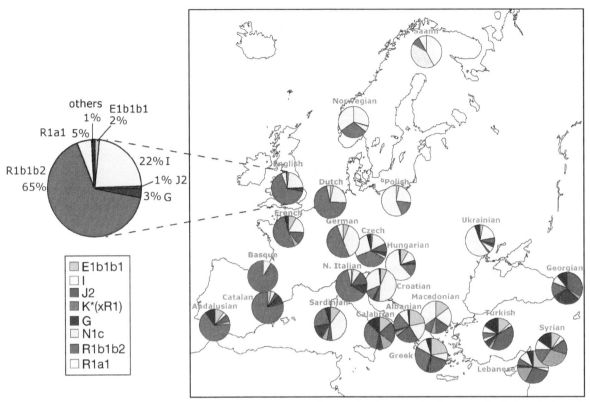

FIGURE 3-8. *Haplogroup diversity in Europe. In Western Europe the dominant haplogroup is R1b1b2 which is the major haplogroup in the British Isles, Norway, Denmark, Sweden, Holland, Germany, France, Spain and, Italy, so by itself is not a very good indicator of ancestry. The proportions for England refer to sampling from eastern England.*

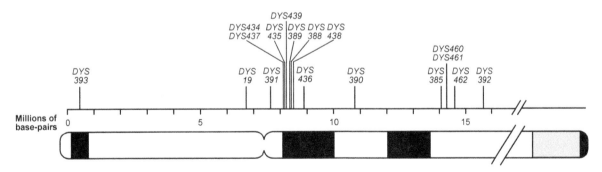

FIGURE 3-9. *The number of copies of tandemly repeated runs of bases at various short tandem repeat (STR) markers (DYS numbers) along the 60,000,000 base pair long Y-chromosome DNA molecule are used to characterize a man's Y-chromosome haplotype. some commonly used markers are indicated.*

from different numbers of repeats of short sequences of bases along the DNA (Figures 3-9, 3-10).

These mutations or polymorphisms, which can change over a time scale of hundreds or thousands of years - as opposed to the more slowly changing SNPs - are referred to as *short tandem repeats "STRs"*, or sometimes *"microsatellites"*. For example, a particular STR known as DYS19 contains a motif of four DNA bases, TAGA, repeated 10-17 times in a man's Y-chromosome. If a man has 14 repeats, this number is then passed on faithfully from generation to generation along the male line, father to son and so on, until, due to an error in the body's genetic replication machinery, the next generation might receive a Y-chromosome with 15, or only 13 repeats. Such changes occur about once in every five hundred generations for each STR, on average. DNA analysis technology known as "PCR" (Polymerase Chain Reaction) now allows the analysis of a set of over 17 of these microsatellite markers (Figures 3-11, 3-12)[81], given names such as DYS19, DYS389I, DYS389II, DYS390, DYS391, DYS392, DYS393, DYS385a, DYS385b, DYS439, DYS458, DYS437, DYS448, DYS456 and DYS438 etc. DYS stands for "<u>D</u>NA <u>Y</u>-chromosome <u>S</u>egment".

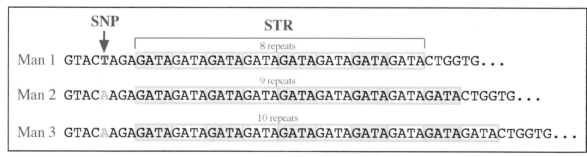

FIGURE 3-10. *Part of the DNA from the Y-chromosomes of three men. Man 1 is different from men 2 and 3 by having a residue called thymine or T – the result of a single nucleotide polymorphism or "SNP" in the position the other males have an adenine A. He is also different from the others because the short tandem repeat or "STR" indicated by the bracket has 8 repeats of the residues GATA whereas man 2 has 9 repeats and man 3 has 10 at that STR. The SNPs along the Y-chromosome DNA specify a man's haplogroup, and the STRs (typically between 6 and 20 of them) define a man's haplotype.*

81. Bosch, E., Lee, A.C., Calafell, F., Arroyo, E., Henneman, P., deKnijff, P. and Jobling, M.A. (2002) High resolution Y-chromosome typing: 19 STRs amplified in three multiplex reactions. *Forensic Science International*, volume 125, pages 42-51.

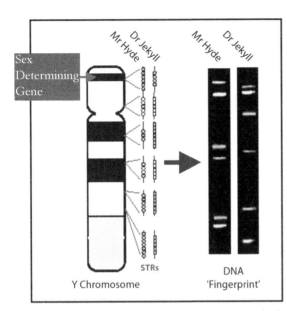

FIGURE 3-11. *Short tandem repeats and the corresponding DNA "fingerprints". After the sample has been collected a man's Y-DNA is analysed in the laboratory. From the positions of the bands appearing on what we call gel electropherograms or simply "gels" a man's Y-chromosome type or "haplotype" can be found in terms of the numbers of repeats or STRs at various locations along his Y-chromosome. In this schematic representation Dr. Jekyll and Mr. Hyde have different haplotypes.*

Six of these STR markers have been widely used by researchers in human population ancestry, and for the Wirral and West Lancashire Viking DNA project these are the ones that were employed: DYS19, DYS388, DYS390, DYS391, DYS392, DYS393. We can now add two more - DYS389I and DYS389II. There are other markers that could have been chosen, but these are not so useful for researching Viking ancestry because (i) the databases do not have extensive data for them, or (ii) they are rapidly changing (mutations occurring in less than 40 generations on average).

Haplotypes reveal finer grained diversity within haplogroups (Figure 3-13), and once a man's haplotype has been determined it is possible to predict his haplogroup, often to a high degree of certainty. Haplotypes provide a much more useful indicator of individual ancestry than haplogroups, using databases such as YHRD and looking for matches.

FIGURE 3-12. *Set of results from a man's Y-chromosome analysis showing his haplotype (in this case for 17 markers) and his haplogroup. His haplotype data can then be entered in to a world-wide database called YHRD ("Y-chromosome haplotype reference database") to look for matches.*

Haplogroup

○ A ○ BC ○ DE ○ F* ○ G ○ H ● I ○ J ○ K* ○ R1* ○ R1a ○ R1b

Haplotype

DYS436	12	DYS391	10	DYS389I	14
DYS437	14	DYS390	23	DYS398II	30
DYS438	10	DYS393	13	461	12
DYS434	12	DYS392	11	462	12
DYS435	11	DYS388	15	460	10
DYS439	12	DYS19	15		

Maternal ancestral line and haplogroups– mitochondrial DNA

Apart from the Y-chromosome, passed from father to son with little or no change, there is an analogous maternally inherited piece of DNA, mitochondrial DNA (mtDNA), which passes from mother to children, and this too can be used to study population movements. Mitochondrial DNA is a much shorter molecule than the non-recombining part of the Y-chromosome. There are no short-tandem repeat sequences, and hence no haplotypes in the sense that we have for the Y chromosome. All variation is due to SNPs; however, the SNP mutation rate is considerably higher than in the Y-chromosome, giving a relatively high degree of diversity. And as with the paternal lineage, we all have a maternal most recent common ancestor (MRCA) which some have called 'mitochondrial Eve'.

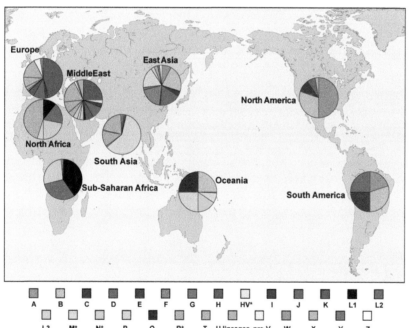

The human mtDNA tree splits into major haplogroups that carry exclusively African sequences and just one haplogroup, L3, that the Africans share with the rest of the world. All non-African mtDNA haplogroups derive from just two lineages (M and N) branching out from the root of haplogroup L3. Common European haplogroups include H, V, J, N, T, U, and W (Figure 3-13).

FIGURE 3-13. *The main mitochondrial DNA haplogroups and their distribution through the world*[82].

82. Copyright 2004 from *Human Evolutionary Genetics* by Jobling, M.A., Hurles, M.E. and Tyler-Smith, C. Reproduced by permission of Taylor and Francis Group, LLC, a division of Informa plc.

FIGURE 3-14. *Volunteer from West Kirby, Wirral, taking the mouth swab test.*

Sampling an individual's DNA

DNA sampling is normally done from a mouth swab of the buccal cells lining the cheek using a specially designed brush[83] (Figure 3-14). Having taken the swab, a volunteer then places it into a small tube containing preservative solution. After a good twizzle the brush is then discarded, the tube is sealed and then taken to the laboratory (Figure 3-15) where the samples are extracted and analysed.

Specially constructed instruments allow the automated analysis of up to 20 DNA markers at a time - this is known as "multiplexing". For a man his Y-chromosome SNP or haplogroup is determined by first of all amplifying the relevant parts of his DNA using a technique known as the polymerase chain reaction, or "PCR" - an excellent web site describing the principles of PCR with an interactive display is:

http://users.ugent.be/~avierstr/principles/pcrani.html

Haplogroups can then be determined from this amplified DNA using a number of methods, one example being what is called the SNaPshot method which determines the sequence of the bases along the amplified regions of the relevant parts of the DNA[84]. A similar procedure can be used to identify a man or woman's mitochondrial (maternal) haplogroup.

---

83. King, T.E., Ballereau, S.J., Schürer, K. and Jobling, M.A. (2006) Genetic signatures of coancestry within surnames. *Current Biology* volume 16, pages 384–388.
84. Bowden, G.R., Balaresque, P., King, T.E., Hansen, Z., Lee, A.C., Pergl-Wilson, G.,Hurley, E., Roberts, S.J., Waite, P., Jesch, J., Jones, A.L.,Thomas, M.G., Harding, S.E. and Jobling, M.A. (2008) Excavating past population structures by surname-based sampling: the genetic legacy of the Vikings in northwest England. *Molecular Biology and Evolution*, volume 25, pages 301-309.

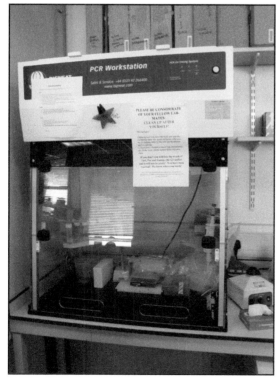

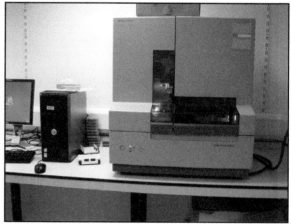

FIGURE 3-15. *Researchers Turi King and Pille Hallast in a modern genetics laboratory (top) and (bottom) some of the equipment used for analysis: (left) area for setting up polymerase chain reactions for amplifying the Y-DNA and (right), automatic analyser for analysing the Y-chromosome STRs and determining a man's Y-DNA haplotype.*

 the

Haplotype analysis

The numbers of short tandem repeats (STRs) for each selected marker (or DNA Y segment, DYS) can also be measured – this allows the Y-haplotype to be determined[85]. As indicated above, this can also usually be used to deduce the man's haplogroup, and a lot of recent research has been focused on predicting haplogroups from haplotypes. A convenient computer programme is available online to do this for a selection of haplogroups: www.hprg.com/hapest5/

Individual ancestry

So, having done the test, how can we comment on a person's ancestry? First of all, for either a man or woman we can see from a European or world map where their maternal haplogroup is most frequently occurring. For example, a haplogroup H mtDNA sequence would normally indicate a European origin, and haplogroup V might imply an origin in the Iberian peninsula. A man can also see where his paternal Y-chromosome haplogroup is occurring, and some lineages that are most frequent in Scandinavia are often taken to indicate individual paternal links to the Vikings.

Individual Viking ancestry from haplogroup and haplotype data

Y-chromosomal haplogroup R1a1 is common in Norway but occurs at much lower frequency in other parts of Western Europe except those regions with known Norse connections. Estimates for modern Norway have ranged from ~35%[86] down to ~26%[87]. Its occurrence in the British Isles is often

85. Jobling, M.A., Hurles, M.E. and Tyler-Smith, C. (2004) *Human Evolutionary Genetics: Origins, Peoples and Disease*. Garland Science Publishing, London/New York, pages 476-480.
86. Capelli, C., Redhead, N., Abernethy, J.K., Gratrix, F., Wilson, J.F., Moen, T., Hervig, T., Richards, M., Stumpf, M.P.H., Underhill, P.A., Bradshaw, P., Shaha, A., Thomas, M.G., Bradman, N. and Goldstein, D.B. (2003) A Y-Chromosome census of the British Isles. *Current. Biology*, volume 13, pages 979-984.
87. Dupuy, B.M., Stenersen, M., Lu, T.T, and Olaisen, B. (2006) Geographical heterogeneity of Y-chromosomal lineages in Norway. *Forensic Science International*, volume 164, pages 10-19.

taken as a sign of Viking ancestry although it is also common across Eastern Europe and India. As we discussed on page 66, Haplogroup N - associated with the Saami people - is also taken as a sign of Norse ancestry, though this is rare in southern Norway and extremely rare in Britain. Haplogroup I1a is also often regarded as a sign of Viking ancestry as it is common across Scandinavia, but is also relatively common in Germany, Holland and Belgium. The commonest haplogroup across Scandinavia is R1b1b2 – unfortunately it is also common across most of Western Europe as a whole. Specific **haplotypes** within I1 and R1b1b2, as well as G and others, have nonetheless been used to give stronger clues to ancestry, using haplotype databases such as YHRD.

The YHRD database

YHRD stands for **Y**-chromosome **H**aplotype **R**eference **D**atabase. Once a man's haplotype has been obtained the DYS values can be entered into the database looking for geographical matches. Some of the DYS markers tend to change or mutate on a timescale of a few generations – whilst these are useful for recent ancestry and paternity testing they are not so useful for ancient ancestry. As noted above on page 69, the following markers have been particularly useful in the pursuit of Viking ancestry: DYS19, DYS390, DYS391, DYS392, DYS393 and one or more of DYS388, DYS389I or DYS389II. Note that matches seen in the YHRD are geographical only, and there is no possibility to contact men carrying matching haplotypes, or to find out what their surnames are.

Individual tests can give "an idea" of personal ancestry

At the haplogroup level all of the Y chromosome types are widespread so it is impossible to be conclusive. As we have already stated, this is especially true of R1b1b2 which, although very common in modern Scandinavia, is also common throughout the rest of Western Europe. For haplogroup I1 one can be a little more confident, although again it is widespread in Northern Europe. For individuals carrying R1a1 we can be more confident, especially if recent ancestry in Eastern Europe – where it is also common – and the

Indian subcontinent can be ruled out. The same is true for haplogroup N if recent north Asian ancestry can be ruled out. These kinds of haplogroup matches can only be regarded as a rough indication.

Haplotypes are more specific and discriminating than haplogroups. This diversity is particularly useful for discriminating within the widespread R1b1b2 haplogroup and several haplotypes have their top matches on a frequency basis in Scandinavia, where the match frequency F^{88} can be expressed as a percentage by multiplying by 100.

Table 3-1. *Some haplotypes within haplogroup R1b1b2 with top matches in Scandinavia. Some haplotypes from I1, I2b1 are included for comparison.*

Table 3-1 shows some of the haplotypes from R1b1b2 with top matches in Scandinavia that were sampled in the Wirral and West Lancashire survey. For example the haplotype with markers DYS19 = 14, DYS390 = 23, DYS391 = 10 , DYS392 = 13, DYS393 = 13, DYS389I = 13 and DYS389II = 29 has a top match on a frequency basis (F ~ 8%) in Jönköpping/Östergötland, in southern Sweden (Figure 3-16a). A blue square means no matches at that place, a square with red in it means matches were found – a full red square means the location with the top number of matches. Table 3-1 also shows for comparison some haplotypes from other haplogroups (I1, G, I2b1).

	Haplotype							
Hg	DYS19	DYS390	DYS391	DYS392	DYS393	DYS389I	DYS389II	Top match
R1b1b2	14	24	10	13	13	13	30	Västerbotten: 5% of men (F~5%)
R1b1b2	15	23	11	13	13	13	31	Southern Norway: F ~1%
R1b1b2	14	23	10	13	13	13	29	Jönköpping/Östergötland: F~8%
I1 and G	14	22	10	11	13	12	28	Gotland: F~15%
I1	14	23	10	11	13	12	28	Jokkmokk: F~21%
I1	15	23	10	11	13	12	30	Blekinge: F~5%
I2b1	15	23	10	12	14	14	32	Västerbotten: F~7%

Hg: Haplogroup; DYS: DNA Y Segment;
F (%) = 100 x (Number of men with a matching Y-haplotype in a particular place or area)/ (Total number of men tested)

88. F(%) = 100 x (Number of men with a matching Y-haplotype in a particular place or area) / (Total number of men tested in that place)

(a) R1b1b2
DYS19 = 14, DYS390 = 23, DYS391 = 10, DYS392 = 13,
DYS393 = 13, DYS389I = 13, DYS389II = 29

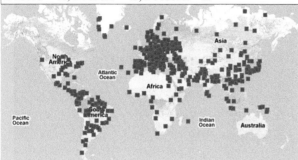

(b) I2b1
DYS19 = 15, DYS390 = 23, DYS391 = 10, DYS392 = 11,
DYS393 = 13, DYS389I = 12 and DYS389II = 30.

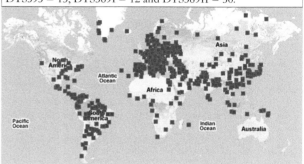

(c) I1
DYS19 = 15, DYS390 = 23, DYS391 = 10, DYS392 = 12,
DYS393 = 14, DYS389I = 14, DYS389II = 32.

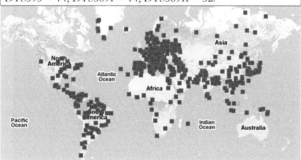

(d) I1
DYS19 = 14, DYS390 = 22, DYS391 = 10, DYS392 = 11,
DYS393 = 13, DYS389I = 12, DYS389II = 28.

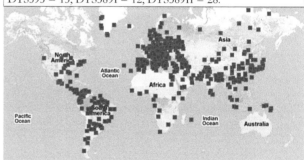

(e) I1
DYS19 = 14, DYS390 = 23, DYS391 = 10, DYS392 = 11,
DYS393 = 13, DYS389I = 12 and DYS389II = 28.

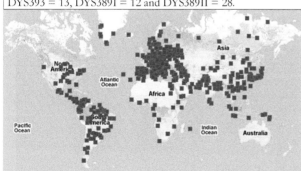

FIGURE 3-16. *Y-chromosome haplotype clusters for 5 men from Wirral and West Lancashire with top matches in a location in Scandinavia.*

Blue squares: no matches in that location. Squares with some red in mean matches found, with a full red square meaning the top matches: the place where their chromosome is most frequently occurring:
a) Jönköpping/Östergötland b) Västerbotten c) Blekinge d) Gotland e) Jokkmokk

Data from the YHRD database yhrd.org

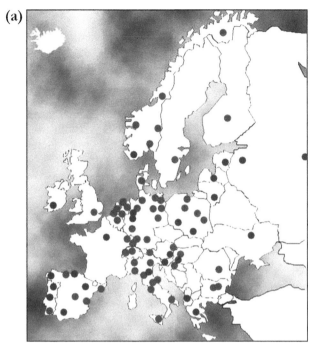

(a)

(b)

Population	Count	Frequency %
Norway Central	3 of 48	6
Norway East	5 of 85	6
Norway Oslo	2 of 33	6
Denmark	4 of 63	6
Norway North	2 of 45	4
Sweden	22 of 510	4
Zeeland	2 of 46	4
Budapest	3 of 117	3
Freiburg	12 of 433	3
Hamburg	3 of 114	3
Latium	6 of 222	3
Norway West	2 of 64	3

FIGURE 3-17. *Matches for the cluster of I1 haplotypes around DYS19 = 14, DYS390 = 22, DYS391 = 10, DYS392 = 11, DYS393 = 13, DYS389I = 12 and DYS389II = 28 for a man from the Wirral, using an earlier version of YHRD (a) blue circles – no matches found; red circles – matches found (b) a list of the places showing the top match frequencies.*

Figure 3-16b shows a haplotype within the I2b1 haplogroup with markers DYS19 = 15, DYS390 = 23, DYS391 = 10 , DYS392 = 12, DYS393 = 14, DYS389I = 14 and DYS389II = 32. These yielded a top match in Västerbotten, Sweden (F=7%). There are a significantly larger proportion of haplotypes with their strongest matches in Scandinavia for the haplogroup I1. For example the haplotype DYS19 = 15, DYS390 = 23, DYS391 = 10, DYS392 = 11, DYS393 = 13, DYS389I = 12 and DYS389II = 30 has a top match on a frequency basis at Blekinge in Sweden (F = 5%) (Fig 3-16c). The haplotype DYS19 = 14, DYS390 = 22, DYS391 = 10, DYS392 = 11, DYS393 = 13, DYS389I = 12 and DYS389II = 28 has a top match on a frequency basis at Gotland in Sweden (F = 15%) (Fig 3-16d) and another I1 haplotype DYS19 = 14, DYS390 = 23, DYS391 = 10, DYS392 = 11, DYS393 = 13, DYS389I = 12 and DYS389II = 28 has a top match on a frequency basis at Jokkmokk in Sweden (F = 21%) and Figure 3-16e shows where matches have been found for this particular haplotype.

In a slightly earlier version of YHRD the results are shown simply as red circles for where there are matches for a man's Y-chromosome haplotype and blue for no matches. A list showing the matches enabled the top matches in terms of the places with the leading match frequencies F to be calculated. An example is shown for a particular haplotype within haplogroup, I1 (Figure 3-17): DYS19 = 15, DYS388 = 12, DYS390 = 22, DYS391 = 10, DYS392 = 11 and DYS393 = 13, also yielded top matches from the YHRD database in Scandinavia with 6% of the male volunteers tested in four regions - Central Norway, East Norway, Oslo and Denmark – showing matches.

In all these cases the individuals whose Y-chromosomes are represented with strong matches in the Viking homelands have other matches over a wide area in Europe. For example the R1b1b2 haplotype DYS19 = 14, DYS390 = 23, DYS391 = 10, DYS392 = 13, DYS393 = 13, DYS389I = 13 and DYS389II = 29 also has strong matches in North Holland and Eastern Europe and the I1 haplotype DYS19 = 14, DYS390 = 23, DYS391 = 10, DYS392 = 11, DYS393 = 13, DYS389I = 12 and DYS389II has strong matches in the Czech Republic and Greece.

What these examples show is that it is impossible to be conclusive with individual data even at the haplotype level: deductions about Viking ancestry at the individual level are therefore *unsafe*, and can only give at best an idea. In addition, not all locations have been equally sampled, so additional future sampling and updates to the YHRD could change the picture for any given haplotype.

Chapter 4
POPULATION ANCESTRY AND SURNAMES

In 1971 the Historian P. Sawyer writing in his book *The Age of the Vikings*[89] had indicated the potential of population genetics as an indicator of population movements, if an appropriate genetic tool could be found. Forty years on the ability to compare **distributions** of Y-chromosome haplogroups and haplotypes has realised the possibility of following male population movements, and the ability to compare distributions of mitochondrial haplogroups has provided the ability to follow the corresponding female population movements.

Such possibilities have helped us address the following two questions concerning the survey:

1.   Is there a significant Norse Viking contribution to the DNA of the modern and old populations of Wirral and West Lancashire?

2.   If there is, can we get an idea of what approximate proportion of the DNA, or "DNA admixture" in the modern and old populations - prior to the Industrial Revolution - in Wirral and West Lancashire is of Norse Viking origin?

Comparing distributions

There are a number of ways of comparing distributions of paternal Y-chromosome DNA haplogroups and haplotypes and maternal mitochondrial DNA haplogroups. These include:

*   Simple inspection of haplogroup frequencies in the form of haplogroup pie charts.

*   Consideration of frequencies of specific haplogroups and haplotypes.

*   Population differentiation tests and genetic distance measures using haplogroup and haplotype data.

*   Population admixture analysis.

89. Sawyer, P.H. (1971) *The Age of the Vikings*, Edward Arnold, London, pages 255-256.

Haplogroup pie charts

We have seen in the previous Chapter how the haplogroup distributions vary across the world and also throughout Europe. This method is good at visualising differences between places but has the disadvantage of not allowing a quantitative comparison and does not allow us to show the uncertainty in estimates for the various groups.

Frequencies of specific haplogroups

Paternal haplogroups R1b1b2 and I1 although very common in Scandinavia are also common in other parts of northern and western Europe so looked at separately are of limited use. R1a1 however is a relatively uncommon haplogroup in Western Europe except in Norway and those areas known to have been settled by Norsemen. Fig 4-1 maps the distribution for Norway, Denmark (Jutland), Holland, and Ireland/central Scotland (representing the indigenous British Isles population). So for the British Isles by simply observing the level of R1a1 in a population we can get a rough *idea* of the possible extent of Norse ancestry in that particular area (Figure 4-1).

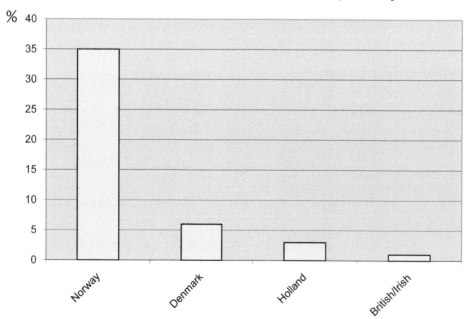

FIGURE 4-1.

Histogram comparing the proportions (%) of the Y-haplogroup R1a1 in modern northwestern European populations.

The comparisons using R1a1 frequencies alone are at best only semi-quantitative and particular care needs to be expressed at excluding the contributions of volunteers deriving from modern Poland, Russia or India for example as these populations also have high levels of occurrence of R1a1.

Population differentiation tests and genetic distance measures

Statistical tests can be carried out that compare the different haplogroups or haplotypes in one population with those in another. We can ask: are these two populations significantly different? – and get a yes or no answer, together with a probability value that the difference could occur just by chance. Statisticians usually think that a difference is significant if it could only arise by chance less than 5% of the time.

Another way to compare populations is to come up with a figure that measures how much 'genetic distance' there is between any pair of populations. These measures are in the form of numbers between 0 and 1, where 0 means no difference between two populations, and 1 means that the populations share no haplogroups/haplotypes at all. We use measures called FST (for haplogroups) and RST (for STR haplotypes). When several populations are compared, we derive a set of distances for each pair, and then these can be plotted on a graph in which each population is a dot, with the distance between dots representing the degree of difference, or genetic distance between them.

Admixture analysis

Population admixture analysis is a way of quantifying the ancestry of a group or population of people. "Admixture" means the relative proportion of DNA in a population coming from different ancestral sources: for example from Norway or from the British Isles population that were present in the northwest of England before the Norse Vikings arrived. A sufficient number of appropriate volunteers representing the ancestral parent populations are required and appropriate controls need to be selected. For our study a procedure developed by Professor Agnar Helgason of the University of Reykjavik was used which employs the haplogroup and haplotype data[90]. The Norwegian parent sample used in the survey (from 201 volunteers) was from a published data set by Dr. Cristian Capelli and coworkers and the indigenous British Isles parental sample (84 volunteers) was assembled from the same source by pooling central Scottish and Irish haplotypes (Pitlochry and Castlerea)[91]. To reduce bias, we equalized the sizes of the parental population samples by randomly selecting only 84 from the larger Norwegian volunteer base and checked with 9 further random selections which showed no significant differences.

Each 'hybrid' population, such as the ones from Wirral or West Lancashire, can then be analysed and given a percentage ancestry from the two 'parental' samples, together with a statistical margin for error.

Criteria for volunteer recruitment

In using genetics to establish ancestral links of groups or populations with the past, suitable criteria for volunteer recruitment have to be established. Modern population movements, particularly since the time of the Industrial Revolution and the onset of modern transport methods - railways, cars, sea and air transport - can dilute any ancestral signal in the DNA of populations. In many recent ancestral studies involving the analysis of the Y- or mitochondrial DNA the usual criterion for volunteer selection is known as a '2-generations of residence' criterion; in other words, to take part in a Y-chromosome survey a (male) volunteer has to be able to say that

---

90. Helgason, A., Hickey, E., Goodacre, S., Vega, E., Bosnes, V., Stefánsson, K., Ward, R., Sykes, B. (2001) mtDNA and the islands of the North Atlantic: Estimating the proportions of Norse and Gaelic ancestry. *American Journal of Human Genetics*, volume 68, pages 723-737.

91. Capelli, C., Redhead, N., Abernethy, J.K., Gratrix, F., Wilson, J.F., Moen, T., Hervig, T., Richards, M., Stumpf, M.P.H., Underhill, P.A., Bradshaw, P., Shaha, A., Thomas, M.G., Bradman, N. and Goldstein, D.B. (2003) A Y-Chromosome census of the British Isles. *Current. Biology*, volume 13, pages 979-984; Goodacre S., Helgason A., Nicholson J., Southam, L., Ferguson, L., Hickey, E., Vega, E., Stefánsson, K., Ward, R. and Sykes, B. (2005) Genetic evidence for a family-based Scandinavian settlement of Shetland and Orkney during the Viking periods, *Heredity* volume 95, pages 129-135.

his paternal grandfather was born in the area. For population comparisons based on mitochondrial DNA (maternal ancestry) the same "2-generation" criterion is applied.

The 2-generation criterion is not usually sufficient for urbanized areas, especially those at or near coastal ports, so either these places are avoided – which excludes several important areas – such as Wirral and West Lancashire, where other evidence has indicated significant Viking settlement – or some additional criterion for volunteer recruitment needs to be used to obviate the effects of modern population movements. For men such a possibility exists. For a male sample we can take advantage of a known strong link between patrilineal surnames and the Y-chromosome. Namely, since approximately the 14th Century in England – well before the Industrial Revolution - men have not only been bequeathing their Y-chromosomal DNA directly down the patrilineal line, but also their surnames. Possession of a surname in the area prior to the Industrial Revolution provides a further criterion for volunteer recruitment. This, however, can only help with the ascertainment of the ancestry of male proportion of populations and cannot help with volunteer recruitment for maternal ancestry.

Genetic studies on the link between surnames and Y-DNA

Though there is evidence of surnames existing in Britain prior to the Conquest in AD 1066, these were not hereditary surnames but rather non-hereditary bynames which changed with each generation and even within a person's lifetime. The practice of hereditary surnames was introduced to Britain by the Normans, who had already been using them for a couple of generations. As in France, the Norman lords in Britain used them as a way of solidifying claims to land and therefore the first surnames were often derived from place-names: e.g. *Rogier de Poitivin*. The use of hereditary surnames was first taken up by the wealthier land-owning families and then trickled down to the rest of the population over the following centuries until by AD 1500 it was becoming rare not to have a surname. Hereditary surnames were adopted earlier in the south than in the north of England and earlier in England than in Scotland and Wales.

The 1881 census lists over 400,000 surnames and, perhaps not surprisingly, most surnames are rare with, on average, under one hundred people carrying the same surname. Many surnames are also highly geographically localized, often being confined to the small region of the country where the surname originated.

In Britain it is the father who passes on his surname to his children and to his male children he also passes on his Y chromosome. Because men inherit their surnames and Y chromosomes together one might expect, therefore, that a link would exist between the two and that that link may have existed for several hundreds of years. Indeed, the Y chromosome type inherited could be that of the surname's original founder. A further expectation arising from this link would be that all men sharing a surname through a common ancestor would all share a Y chromosome type.

There are a number of factors that would confound the simple link between a surname and a Y chromosome type. Some surnames were adopted independently by more than one individual at the time of surname foundation and this will have resulted in more than one Y chromosome type being associated with a single surname. The link between and surname and a Y chromosome type will be broken by any event where the two are not inherited together: illegitimacy, adoptions or a deliberate surname change result in men having one man's surname but another man's Y chromosome type. As stated above, mutation can cause slight changes to occur on the Y chromosome as it is passed from generation to generation and could result in small haplotype differences between men who share a common ancestor through their surname. However, we know the mutation rates of many Y chromosome markers, and this can therefore be taken into account when assessing the nature of the differences between Y chromosome types: SNP mutation rates are so low, and surname histories so short, that we don't expect to see differences between men at these markers. On the other hand, STRs mutate much more frequently and therefore differences between men who share a common ancestor through their surname are more likely to be observed. Finally, 'genetic drift', the random differences in the number of male offspring that men have, will affect the number of the original Y chromosome lineages tied to a surname that survive to the present day: some surname lineages will have died out whereas others will have flourished.

So what of this link between surname and Y chromosome? Does it exist and why are we interested in using it to explore the genetic legacy of the Vikings? A link with a surname provides us with the means to tie a Y chromosome to the past. We can use information about surnames to allow us to sample individuals who are more likely to have Y chromosome types associated with a certain region several hundred years ago: for example, by selecting men with surnames known to have been established long ago, and confined to particular parts of the country.

There are two main ways to assess the Y chromosome/surname link. The first method assesses the general link by simply taking two men geographically randomly who share the same surname and testing to see if their Y chromosomes are identical or nearly identical. If surnames and Y chromosomes are faithfully passed down together then we would expect men who share a surname to share a Y chromosome type. While we would not expect this to be so for common surnames, where more than one man probably founded the surname, for rarer surnames, a link could exist. A recent large study of this kind[92] has found that such a link does indeed exist with the relationship being stronger the rarer the name.

The other method is to look at single surnames in depth, collecting a large number of seemingly unrelated men who share the same surname. The first study of this kind examined the surname *Sykes* and while a link was found, the study was of low resolution using few STR markers. Much larger studies at higher resolution examining many different surnames have since been carried out both in Britain and Ireland. In Britain, the study recapitulated the finding of the more general study above: among men with a high frequency name, such as Smith, there is little sharing of Y chromosome type. However, the amount of haplotype sharing occurring among men who carry a rarer surname can be striking: for example, among randomly recruited, seemingly unrelated, men who share the surname Attenborough, 87% share an identical or near-identical Y chromosome type. In Ireland it was found that the link existed regardless of the frequency of the surname: even some very common surnames showed a strong relationship between surname and Y chromosome type.

92. King, T.E., Ballereau, S.J., Schürer, K. and Jobling, M.A. (2006) Genetic signatures of coancestry within surnames. *Current Biology* volume 16, pages 384–388.

A convenient way of showing the relationship of Y-haplotypes within a surname is the use of network diagrams. Figure 4-2a is a schematic diagram showing haplotypes represented by circles, the colours denoting which haplogroup they belong to. The area of the circles are proportional to the frequency of the haplotype within the dataset. Lines represent mutational steps with the shortest line representing a single mutational step. Fig 4-2b shows an actual example for Y-haplotypes of men bearing the name *Swindlehurst*, a name which appears to originate from the northwest of England.

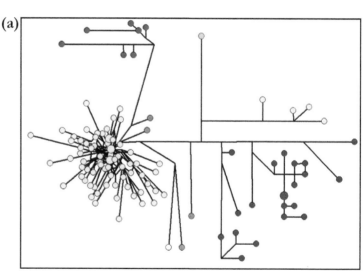

FIGURE 4-2. *Network diagrams show how close, or far apart, Y-haplotypes are to or from each other. Circles represent Y haplotypes with the area of circles being proportional to the frequency of the haplotype within the dataset. Lines represent mutational steps with the shortest line representing a single mutational step, and colours represent different haplogroups. (a) Schematic example (b) an example for the surname Swindlehurst. Red circles are different haplotypes within haplogroup R1a1, yellow circles R1b1b2, blue within I1, and turquoise within R1*. The larger the circles the more Swindlehursts with that haplotype.*

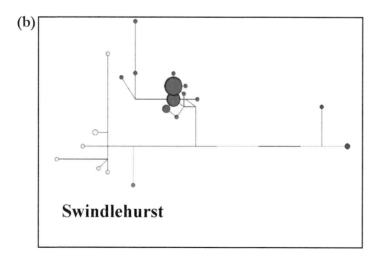

From these studies it is clear that, certainly for rarer surnames, a link between surname and Y chromosome exists and therefore, ties a Y chromosome type to the past through the surname. We are interested in knowing about the types of Y chromosomes that were present in a region following the Viking period. However, recent population growth and movement, such as that during the Industrial Revolution, will have blurred the signal of the Viking migration. Surnames are some 700 years old and contain within them information about where they originated and therefore allow us to leap-frog over the effects of more recent population movement. We can use surnames from old lists of people paying taxes, criminal records and church records as a way of sampling modern individuals and looking at the types of Y chromosomes associated with a region several hundred years ago. In this way we can match old DNA with modern geography, and we have used this strategy in the present study.

Chapter 5
EARLIER STUDIES ON THE BRITISH ISLES

The first systematic study of ancestry for selected regions of the British Isles was conducted by Cristian Capelli and coworkers and was published in Current Biology in 2003, based on the male Y-chromosome[93]. This was followed by a series of papers by Agnar Helgason, Sarah Goodacre and coworkers who looked at the comparative Scandinavian ancestry of male and female populations based on Y-chromosome and mitochondrial DNA distributions for the Orkneys/Shetlands, the Western Isles and Isle of Man (as well as Iceland and the Faroe Islands)[94]. All of these studies were based on a 2-generations of residence criterion for volunteer recruitment. Finally McEvoy and colleagues in 2006[95] published a Y-chromosome survey of Ireland using surnames to aid volunteer recruitment but found no evidence of Viking ancestry. We consider each of these studies in turn before we describe the results in the following Chapter from the Wirral and West Lancashire survey.

The study by Capelli and co-workers (2003)

A team based at University College London conducted a survey of the British Isles, involving 1772 men from predominantly 25 small urban locations in the British Isles: Shetlands, Orkney, Durness (north coast of Scotland), the Western Isles, Stonehaven, Pilochry, Oban, Morpeth, Penrith, Isle of Man, York, Southwell, Uttoxeter, Llanidloes, Llangefni (Anglesey), Rush (Dublin), Castlerea (Ireland), Norfolk, Haverfordwest, Chippenham,

93. Capelli, C., Redhead, N., Abernethy, J.K., Gratrix, F., Wilson, J.F., Moen, T., Hervig, T., Richards, M., Stumpf, M.P.H., Underhill, P.A., Bradshaw, P., Shaha, A., Thomas, M.G., Bradman, N. and Goldstein, D.B. (2003) A Y-Chromosome census of the British Isles. *Current Biology*, volume 13, pages 979-984.
94. Goodacre S., Helgason A., Nicholson, J., Southam, L., Ferguson, L., Hickey, E., Vega, E., Stefánsson, K., Ward, R. and Sykes, B. (2005) Genetic evidence for a family-based Scandinavian settlement of Shetland and Orkney during the Viking periods. *Heredity* volume 95, pages 129-135.
95. McEvoy, B., Brady C., Moore, L.T. and Bradley, D.G. (2006) The scale and nature of Viking settlement in Ireland from Y-chromosome admixture analysis. *European Journal of Human Genetics*, volume 14, pages 1288-1294.

Faversham, Midhurst, Dorchester, Cornwall and the Channel Islands. They compared the distributions of Y-chromosome haplogroups and also haplotypes based on 6 STR markers (DYS19, DYS388, DYS390, DYS391, DYS392, DYS393) for these places with each other and also Norway (using samples from Trondheim and Bergen - representing Norwegian Vikings), Schleswig-Holstein (representing Anglo-Saxons) and Denmark (Danish Vikings). Castlerea was chosen as representing the indigenous British Isles population. For the samples in the British Isles the 2-generations criterion was applied (volunteers had to confirm their paternal grandfather was born at that place). The results showed clear differences within different areas of the British Isles with significant Norwegian input into the populations of Orkney, Shetlands, Durness, Western Isles, Isle of Man and Penrith. The Danish and Anglo-Saxon controls proved indistinguishable although very different from Castlerea. The Capelli team collectively called the Anglo-Saxon and Danish signal as "*Invaders*" and found the Invader signal was generally only significant in those areas of England which also had a significant density of Scandinavian place names (e.g. areas in the former English Danelaw). This implied that the overall contribution of the Anglo-Saxon signal to the genetic admixture of the British Isles was only small, and when the invader signal was strong this was probably due to Danish Vikings.

The study by Goodacre, Helgason and co-workers (2005)

Sarah Goodacre, now at the University of Nottingham, with Agnar Helgason of the University of Reykjavik, and coworkers, attempted to extend the male based study of Capelli *et al.* by comparing the different contributions of Scandinavians to the genetic admixture of the male and female populations in the North Atlantic region, based on analysis of respectively Y-chromosomal DNA and mitochondrial DNA genetic markers. Their findings, based on a study of male and female volunteers from Shetlands, Orkneys, the Western Scottish Isles and Iceland suggested that while areas close to Scandinavia, such as Orkney and Shetland, may have been settled primarily by Scandinavian family groups, by contrast lone Scandinavian males - who later established families with female subjects from the British Isles - may have been prominent in areas more distant from their homeland: the greatest bias towards Scandinavian males was in Iceland (Figure 5-1).

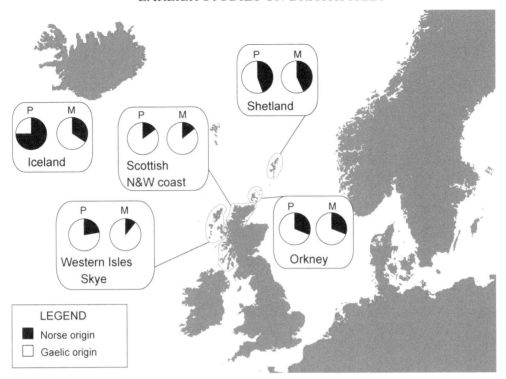

FIGURE 5-1. *Genetic admixture of male and female populations in the North Atlantic[96]. The admixture results indicate that the further away from the Scandinavian homelands men settled the less likely they would take their families with them.*

The study by McEvoy and coworkers (2006)

The Capelli study found little evidence of Viking ancestry in Ireland, perhaps surprising as the Vikings played a considerable role in Irish history founding major places like Dublin and Waterford. In a follow-up study Brian McEvoy, Dan Bradley and colleagues undertook another genetic survey of Irish males but focusing on those who had putative Norse surnames. Although the incidence of such putative Norse names is low, a sample population of 47 men (26 different names but with some surname duplication) was recruited with the following names (numbers in brackets refer to the numbers of unrelated volunteers with that name): Arthur (2), Beirne (2), Bligh (1), Boland (3), Caskey (1), Coll (1), Coppinger (2), Doyle (4), Gohery (2), Hanrick (2), Harold (1), Hendrick (1), Higgins (2), Kells

---

96. From Goodacre, S., Helgason, A., Nicholson, J., Southam, L., Ferguson, L., Hickey, E., Vega, E., Stefansson, K., Ward, R. and Sykes, B. (2005) Genetic evidence for a family-based Scandinavian settlement of Shetland and Orkney during the Viking periods. *Heredity*, volume 95, pages 129-135. Reprinted by permission from Macmillan Publishers Ltd, © 2005.

(2), Kettle (1), Loughlin (1), McGetrick (3), McLoughlin (3), Nelson (1), Norris (1), O'Higgins (1), Sugrue (2), Sweetnam (1), Thunder (1), Toner (3) and Tormey (3). Most of these came from geographic areas associated with Norse activity in Ireland. Admixture proportion estimates in this putative Norse surname group were highly consistent and showed no trace of Scandinavian ancestry.

Despite the wealth of archaeology particularly in Dublin and the finds at the former settlement at Wood Quay, the rarity of Norse names – and rarity of place names (including field names) – contrasting strongly with the Isle of Man and coastal regions of North West England – are consistent with the finds of McEvoy and colleagues and suggest the Norse incursions into Ireland between the 9th and 11th Centuries may have been transient and the mass expulsions of AD 902 (Ingimund's story) and 1016 (Battle of Clontarf) may have been more or less complete.

So in summary of these earlier studies

- In terms of Y-chromosome distributions in populations, modern Norway appears different from other West European populations in having a prevalence of the haplogroup R1a1. In the British Isles evidence of significant Norwegian Viking ancestry has been found in Shetlands, Orkneys, the north coast of Scotland, Western Isles, Isle of Man and the area around Penrith.

- Y-chromosome distributions of modern Denmark and areas of northern Germany and Friesland associated with the Anglo Saxon homelands are indistinguishable, but are very different from Ireland, Scotland and Wales. In England, "Invader" ancestry is generally high in the former Danelaw region found where the density of Scandinavian place names is high – and is generally low elsewhere

- No genetic traces are left of Vikings in Ireland, consistent with little place-name evidence.

Chapter 6
WIRRAL AND WEST LANCASHIRE SURVEY: VOLUNTEER RECRUITMENT

The environs surrounding the Port of Liverpool - the Wirral peninsula and West Lancashire - have undergone a significant flux and expansion in population since the huge expansion of the port following the start of the 19th Century. Taking Wirral as an example, the population at the 1801 census was approximately 10,000. The population today is now approaching 400,000, a population increase in 200 years of 40-fold: this is something like six times the national average increase. This means that for the Survey it was important to adopt a rigorous volunteer recruitment strategy that would get round the influx of people from elsewhere that has been largely responsible for this population increase. In order to get the clearest possible picture of the genetic make up of medieval Wirral and West Lancashire it is necessary to focus on the Y-chromosomes of people bearing surnames that were present in the region either in or shortly after the medieval period.

As we considered in Chapter 4, surnames in the form that we know them now were not passed from father to son in early medieval times: it was not until the 14th-Century that this became commonplace, and this practice was still not followed in some areas. However, we can make the reasonable assumption that the existence of a surname in the 14th - 16th Centuries which corresponds to anything other than the name of a place outside the region, is reflecting someone whose family has been there for many generations. And if we make the additional assumption that there has not been extensive illegitimacy in surname lines from this period to the present then with the criterion of someone possessing an old Wirral or West Lancashire name, we can focus on a much more representative set of Y-chromosomes for the region than just using the standard 2-generations paternal grandfather criterion. A similar surname criterion was successfully applied by J. Wilson and D.B. Goldstein in a survey of the Orkneys[97].

---

97. Wilson, J. and Goldstein, D.B. (2000) Correlation between Y chromosomes and surnames in Orkney, *American Journal of Human Genetics*, volume 67, pages 926–935.

FIGURE 6-1. *Henry VIII - his subsidy rolls of 1542 helped to provide a surname base for volunteer recruitment.*

Modern and 'Medieval' populations

For the Wirral and West Lancashire Viking DNA project two independent samples of unrelated males representing the Wirral, and two representing West Lancashire, were collected using different criteria. The first type of sample set (which we refer to as 'modern') is based simply on two generations of residence, and ignores surname. The second sample set (which we refer to as 'medieval'), was collected using both a residence criterion, and surname information. Subjects were required to have at least two generations of residence, and to have their earliest recorded patrilineal ancestor born in the relevant area. Also, subjects were required to carry surnames that were present in the relevant region prior to 1572 as judged by documentary sources.

Old Wirral and West Lancashire surnames

For Wirral the Henry VIII (Figure 6-1) subsidy rolls recording all households paying taxes in 1542 provided us with a list of 238 old surnames[98]. This list (given as their modern forms) is as follows:

> Adam, Allin, Alleyne, Andrew, Aspinall, Ball, Barber, Barker, Barrell, Barrow, Bailiff, Beck, Bennett, Bergs, Billing, Bird, Blackburne, Boland, Brant, Bratherton, Browne, Brunt, Burscough, Bryde, Burrows, Bushell, Caley, Carr, Carlile, Carlisle, Challoner, Charnock,

---

98. Roberts, S.J. (2002) *A History of Wirral.* Phillimore, Chichester, UK.

Chantrell, Coley, Colley, Colton, Coke, Corf, Corfe, Corness, Cotton, Cowper, Cross, Dalby, Dane, Danold, Davey, Davy, Denham, Denson, Dobb, Doe, Done, Duke, Dunn, Edmonds, Edmunds, Ellcock, Fazackerley, Fiddler, Fidler, Foreshaw, Forshaw, Fox, Francis, Gallie, Gardener, Gardiner, Gardner, Garratt, Garrett, Gibson, Gill, Gleave, Glegg, Goodacre, Grace, Gray, Gregory, Green, Grey, Grice, Hale, Hancock, Hand, Hare, Harper, Harrison, Harvey, Heath, Helsby, Hender, Hesketh, Hey, Heyward, Hide, Hill, Hogg, Hole, Holme, Holmes, Home, Hough, Hulme, Hulmes, Humphrey, Huntington, Hynes, Jennion, Jensen, Jeunds, Johnson, Jump, Kemp, Kirk, Kirkby, Lancelyn, Leck, Ledsham, Leighton, Lennard, Leonard, Ley, Lightfoot, Linacre, Little, Lunt, Macklin, Massie, Massey, Matthew, Mayle, Mayles, Middleton, Milner, Molyneuz, Moss, Moulding, Mutton, Nelson, Newbold, Newton, Otter, Otty, Page, Parr, Pearson, Pemberton, Pendleton, Pennington, Penketh, Penney, Philip, Phylip, Pigot, Pinnington, Plumbe, Poole, Potter, Prenton, Pye, Pyke, Radcliffe, Rathbone, Ravenscroft, Richardson, Rider, Ridley, Rimmer, Robinson, Rogerson, Russell, Rutter, Saddler, Sadler, Sampson, Scaife, Scarff, Scarffe, Scarisbrick, Sclater, Scriven, Sefton, Sharpe, Shephard, Shepherd, Sherlock, Skinner, Smalley, Smythe, Spenser, Stones, Swain, Swaine, Swarbrick, Swindley, Tarleton, Taskar, Tellett, Thomason, Thomasson, Thomson, Threadgill, Threadgold, Tottey, Totty, Tumath, Tyldesley, Wade, Wainwright, Walley, Walton, Warburton, Waring, Warington, Watmough, Watt, Whalley, Wharton, Wilkinson, Williamson, Whitby, Whitehead, Whitelaw, Whitfield, Whitmore, Whittle, Whyte, Williamson, Willoughby, Worral, Woods, Woodward, Wilcock, Wise, Wyse, Young, Yoxon.

To this list could be added a further 2 names from criminal proceedings of 1353[99] (Poole and Harding) (Figure 6-2) and one from ale-house licensing records from 1572[100] (Joynson). This was supplemented further by surnames

---

99. Booth, P. (1983) Calendar of Cheshire Trailbaston Proceedings 1353. *Cheshire History* volume 12, pages 24–28.

100. Bennett, J.H.E. and Dewhurst, J.C. (1940) Quarter sessions records with other records of the Justices of the Peace for the County Palatine of Chester 1559-1760 together with a few earlier miscellaneous records deposited with the Cheshire County Council. *Records Society of Lancashire and Cheshire* volume 94, pages 37–39.

based on specific old local place names such as Upton, Raby and Helsby.

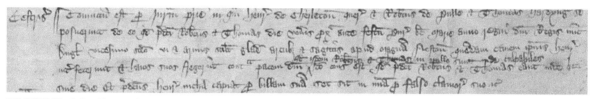

FIGURE 6-2. *Criminal proceedings from the Chester Trailbaston Proceedings of 1353: A transcription of the piece (courtesy of Dr. Paul Cavill) includes the following: Henry Cherleton complained that Robert Poole and Thomas Harding killed his dog at Great Neston, Wirral on Friday 1st Feb. 1348 and broke his hedges. They denied guilt. Jury verdict - Not guilty. Courtesy of the National Archives, Kew, Surrey.*

For West Lancashire, we could call on a list of 232 surnames based on a list of inhabitants of Ormskirk, Scarisbrick-with-Hurlton, Bickerstaffe, Burscough-with-Marton, Westhead-with-Lathom and Skelmersdale who promised to contribute to the stipend of the priest of the altar of Our Lady at Ormskirk in 1366. The document, at the Lancashire Record Office, Preston, was reproduced in the *Ormskirk & District Family Historian*, Spring 1991 (ISBN 0 947915 28 1)[101]. The transcription to modern day names (shown in brackets) was done by Stephen Roberts[102]:

> *Bakhous* (Backhouse), *Balsagh* (Balshaw), *Barett* (Barret-t), *Benyood*, *Bere* (Bere/Bear etc.), *Beyson* (Benson?), *Blanchard* (Blanchard), *Bron* (Brown), *Byld* (Bold), *Byrd* (Bird), *Cadyk* (Cadick), *Carles* (Charles?), *Cauns, Cay,* (Kay), *Childesfadre* (Childsfather), *Coly* (Coly), *Davy* (Davy), *de Adburgam,* de *Aghton* (Aughton), *de Asshurst* (Ashurst), *de Aykescogh* (Aikscough), *de Balshagh* (Balshaw), *de Barton* (Barton), *de Beulond, de Blyth* (Blythe), *de Blythe* (Blythe, Bligh etc.), *de Boold* (Bold), *de Bretherton* (?) (Bretherton/Brotherton?), *de Bronburgh* (Bromborough), *de Bronylegh* (Bronley, and variants), *de Burscogh* (Burscough), *de Bykerstath* (Bickerstaff), *de Cole* (Cole), *de Couper* (Cooper), *de Depdale* (Deepdale), *de Dewicar* (Dewacre), *de Eggeacr*

101. Peet, G. (1991) Inhabitants of Ormskirk, Scarisbrick with Hurlton, Bickerstaffe, Burscough with Marton, Westhead with Lathom and Skelmersdale who promised to contribute to the stipend of the priest of the altar of Our Lady at Ormskirk, 1366. *Ormskirk District Family Historian* volume 1, pages 2–6.
102. The Queen Katherine School, Appleby Road, Kendal, Cumbria.

(Eggacre), *de Ellerbek* (Ellerbeck), *de ffourokshagh*, *de Fletcher* (Fletcher), *de Goldicar* (Goodacre), *de Gosfordsich*, *de Greceby* (Greasby), *de Halsagh* (Halshaw), *de Hamelton* (Hamilton), *de Haskeen* (Haskin(s)/Askin(s)/ Astin/Ashken/ Haskings etc.), *de Haylegh*, *de Holand* (Holland), *de Holbrok* (Holbrook), *de Horscar* (Horsecarr), *de Hurlton* (Hurlton), *de Hyllome*, *de Hyton* (Huyton), *de Irby* (Irby), de Kirkeby (Kir(k)by), *de Ledebeter* (Leadbetter), *de Legh* (Leigh/Lee/Lea), *de Leyland* (Leyland), *de Longebak* (Longback), *de Longeton* (Longton), *de Marton* (Marton), *de Mell* (Mell), *de Mellyng* (Melling), *de Milner* (Milner), *de Morcroft* (Moorcroft), *de Mosbury* (Mossbury), *de Moscar* (Mossock), *de Mosok* (Mossock), *de Mourehyles* (Moorhills), *de Oldome* (Oldham), *de Olton* (Olton/Oulton), *de Orell* (Orrell), *de Owatton* (Overton), *de Owynbrek* (Owenbreck?, Overbeck), *de Par* (Parr), *de Penwytham* (Penwortham), *de Prestcotte* (Prescot), *de Raynford* (Rainford), *de Raynhull* (Rainhill), *de Ruynacre*, *de Shirwallaacrs*, *de Stryvelyn*, *de Sutton* (Sutton), *de Tailour* (Taylor), *de Teulond de Tildeslegh* (Tyldesleigh), *de Westheved* (Westhead), *de Wynmarleigh* (Winmarleigh), *de Wyresdale* (Wyresdale), *del Abbay* (Abbey), *del Aspynwall* (Aspinall/Aspinwall/Astmole), *del Bakhous*, *del Barwe* (Barrow), *del Brodfeld* (?) (Broadfield), *del Brodheved* (Broadhead), *del Brokefeld* (Brookfield), *del Car* (Carr), *del Crosse* (Cross), *del Grene* (Green), *del Greves* (Greaves), *del Halle* (Hall), *del Helmes* (Holmes), *del Heth* (Heath), *del Hyles*, *del Lone* (Lone?), *del Marhalgh* (Marhall?), *del Mor* (More/Moore), *del Mosse* (Moss), *del Outsich* (?), *del Platt*, (Platt), *del Rydyng* (Riding), *del Scoles* (Scholes/Scales), *del Shagh* (Shaw), *del Strenger* (Stranger), *del Syche* (Such/Sutch/Souch/Zouch/ Chuck/Chucks), *del Tou*, *del Wall* (Wall), *del Warinawro*, *del Wodes* (Woods), *del Wolfall* (Wo(o)lfall), *del Yate* (Yate-s), Dobbeson (Dobson), Doggeson (Dodgeson), Drake (Drake), Dykounson (Dicconson/ Dickinson), *Ekirgarth*, *Elot* (Elliott), *fe Copphull* (Copphull, Coppell), *fel Vale* (Vale), *ffaber* (Smith), *Ffarwys*, *Ffox* (Fox), *fi Robti* (Roberts/ Robertson), *fil Alex* (Alexson?), *fil Beco*, *fil Carpent* (Carpenter), *fil Elyn* (Ellin), *fil ffabri* (Smithson), *fil Hankok* (Hancock), *fil Henr* (Henry-son), *fil Johis* (Jones-son), *fil Johis Becokson* (Becokson), *fil Maryot* (Marriott), *fil Nicholl* (Nicholson), *fil Thom* (Thomas/ Thompson etc.), *fil Walt'i* (Walterson), *fil Willi* (Williams-son), *filia Boold* (Bold-son), *fillia Nutricus* (Foster-son), *Foluler*(?) (Fuller?), *Garc Thom* (Thomas-man), *Gillseson* (Gilson), *Gray* (Grey/Gray),

Hawot, *Hennson* (Henson), *Hertblod* (Hartblood), *Hopcrone*, *Horbert* (Hubert?), *Jakson* (Jackson), *Johis* (Jones), *Kemp* (Kemp), *Lagard* (Lagard), *le Bagger* (Badger), *le Bakster* (Baxter), *le Barker* (Barker), *le Blawer* (Blower), *le Bower* (Bower), *le Brabayn* (Brabourn), *le Breton* (Bre(re)ton), *le Cart* (Carter), *le Chalonner* (Chalonner/Challener/ Challender, Challenor etc.), *le Clerk* (Clerk), *le Coudrey* (Cowdrey/ Cowdray/Cowdroy etc.), *le Cropper*, (Cropper), *le ffreshe* (Fresh), *le Halleknave* (Hallknave), *le Hunt* (Hunt), *le Kekker* (Checker), *le Kyng* (King), *le Lauder* (Lauder), *le Long* (Long), *le Mercer* (Mercer), *le Parker* (Parker), *le Porter* (Porter), *le Prestesmon* (Priestman), *le Reder* (Reader?), *le Salter* (Salter), *le Scuster* (Shuster), *le Sergeant* (Serjeant/Sargeant etc.), *le Smyth* (Smith), *le Spenc* (Spence), *le Spencer* (Spencer), *le Spicer* (Spicer), *le Sporier* (Spurrier), *le Swoon* (Swan), *le Tasker* (Tasker, Taskar), *le Turnour* (Turner), *le Walker* (Walker), *le Walsh* (Walsh), *le Warner* (Warner), *le Webster* (Webster), *Lenne*, *Mabbeson*, *Madok* (Madoc-k), *Materei*, *Messenger* (Messenger), *Milde*, *Moubyn* (Maulby?), *Nykson* (Nickson), *Oety* (Otti/Oti/Otty), *Olyf* (Ollif), *Owasen* (Owenson?), *Owitheved* (Outhead?), *Pacok* (Peacock?), *Page* (Page), *Parlement* (Parliament), *Paweson* (Pawson), *Penydale* (Penidale/Pennydale), *Pety* (Petty), *Pye* (Pye/Pie), *Pygyn* (Pigeon), *Pykhare* (Pickhare?), *Relict. Ade Osmarshagh*, *Robynson* (Robinson), *Ryout* (Rideout), *Ryvelyng*, *Shakelauedy* (Shakalady), *Shakerewet*, *Smalshagh* (Smallshaw), *Sowerby*, *Spurwyn Spynk* (Spink-s), *Staynes* (Staynes), *Steell* (Steele), *Stotfoldshagh* (Stotfoldshaw), *Stykk*, *Tabart* (Tarbert), *Tewe* (Tew), *Thomasmon* (Thomasson), *Todd* (Todd), *Toppyng* (Topping), *Tynkeler*, *Tysing*, *Waryng* (Waring), *Willison* (Wilson), *Wodeloft* (Woodloft), *Wyld* (sic) (Wild), *Wyldebold* (Wildbold/Wildblood), *Wynmenske*.

As with Wirral this list this was supplemented with surnames based on specific old local place-names, and allowing additional surnames containing particular old place-name elements.

Volunteer recruitment: modern population

Recruitment of volunteers representing the modern Wirral and West Lancashire populations – men whose paternal grandfathers were from either of these areas but without an "old surname" requirement was done by sampling in or outside shopping centres/precincts in 5 defined test areas: West Kirby, Woodchurch/Thingwall and Ellesmere Port in Wirral and Ormskirk and Skelmersdale in West Lancashire. This was supplemented by volunteers recruited in conjunction with local dentists. Men whose paternal grandfathers were from the heavily industrialized areas of Birkenhead/Tranmere and Liverpool were not included. 100 volunteers from Wirral and 49 from West Lancashire were recruited in this fashion.

Volunteer recruitment: old or 'medieval' population

An independent set of volunteers was recruited representing the old or 'medieval' population. Volunteers were required to have their paternal grandfather born in either Wirral or West Lancashire and to have their earliest recorded patrilineal ancestor born in the relevant area. Volunteers were also required to have a surname present in the relevant region prior to 1572 as judged by the lists given above. Volunteer recruitment representing "old" Wirral and "old" West Lancashire was made possible with the help of local organisations such as the West Lancashire Heritage Association and Wirral Council, BBC Radio Lancashire, BBC Radio Merseyside, the Liverpool Daily Post, Wirral News, Wirral Globe, Ormskirk Advertiser and the Ormskirk Champion. Special recruitment sessions were organised with the help of Dr. David Favager and Wirral Grammar School and Mr. Patrick Waite and the West Lancashire Heritage Association (Figure 6-3).

FIGURE 6-3.

Volunteers being sampled at a special session organized by the West Lancashire Heritage Association at Ormskirk on 13th November 2002.

The latter event – held at Hurlston Hall, Ormskirk held on 13th November 2002 coincided with the 1000th anniversary of St. Brice's day massacre in AD 1002 when King Æthelred (the unready) 'ordered to be slain all the Danish men who were in England'[103].

Within each of the sample groups for old Wirral and old West Lancashire, duplicate surnames (including known spelling variants) were avoided, though some surnames are found in both Wirral and West Lancashire samples, carried by different and, to our knowledge, unrelated individuals. Using this approach, 37 and 40 males were recruited respectively for old or medieval Wirral and old West Lancashire. The surnames of the recruits were as follows:

Wirral: Barker, Beck, Bennett, Billing, Bird, Bryde, Bushell, Colley, Corfe, Edmunds, Forshaw, Gill, Green, Harding, Hesketh, Holmes, Hough, Joynson, Kemp, Kirk, Lunt, Oxton, Raby, Rathbone, Richardson, Rimmer, Robinson, Sampson, Scarisbrick, Sherlock, Skinner, Taskar, Tellett, Tottey, Upton, Young.

West Lancashire: Alker, Balshaw, Bilsborrow, Brown, Carr, Charnock, Coly, Cook, Cooper, Corfe, Crombleholme, Fletcher, Gill, Gray, Hesketh, Holland, Holmes, Hulme, Jones(son), Leyland, Lunt, Melling, Molyneux, Otty, Pendleton, Penketh, Pennington, Prescott, Rigby, Rimmer, Risley, Roby, Scarisbrick, Sephton, Serjeant, Swarbrick, Thomason, Walsh, Webster, Westhead.

---

103. An action which prompted the Danes under King Svein Forkbeard to subsequently attack and take control of England, leaving England under Danish Rule until 1042.

Chapter 7
THE RESULTS

The Y-chromosome haplogroups and haplotypes were obtained for all the volunteers from the four sets of populations, namely the two modern sets, called "Modern Wirral" and "Modern West Lancashire" (based on a two generation only criterion – paternal grandfather from that area), and the two 'medieval' sets called "Medieval Wirral" and "Medieval West Lancashire" (based on a two generation criterion - together with the old surname criterion).

The haplogroup/haplotype distributions for the modern Wirral and West Lancashire populations were compared with existing[104] 2-generation data for nearby populations on the western side of Britain (the Isle of Man, Llangefni and Penrith), and also mid-Cheshire. The latter was chosen because of its close proximity to Wirral and West Lancashire, but with a much lower proportion of Scandinavian placenames. A sample ('British/ Irish') representing the putative population of the western British Isles prior to the Viking incursions was made by pooling central Scottish and central Irish samples. Samples from Orkney and Shetland were included since these are known to show substantial Scandinavian admixture[105], and a sample from modern Norway was used to represent a possible Viking source population.

The change in the distributions for the medieval Wirral and West Lancashire samples with respect to their modern counterparts was then examined and then estimates of the Scandinavian admixture of the populations were made by further comparison with haplogroup/haplotype distribution data for Norway.

104. Capelli, C., Redhead, N., Abernethy, J.K., Gratrix, F., Wilson, J.F., Moen, T., Hervig, T., Richards, M., Stumpf, M.P.H., Underhill, P.A., Bradshaw, P., Shaha,A., Thomas, M.G., Bradman, N. and Goldstein, D.B. (2003) A Y-Chromosome census of the British Isles. *Current Biology*, volume 13, pages 979-984.
105. Goodacre S., Helgason A., Nicholson, J., Southam, L., Ferguson, L., Hickey, E., Vega, E., Stefánsson, K., Ward, R. and Sykes, B. (2005) Genetic evidence for a family-based Scandinavian settlement of Shetland and Orkney during the Viking periods. *Heredity*, volume 95, pages 129-135.

Haplogroup distributions

Haplogroup analysis of the Y chromosomes of the 'modern' and 'medieval' Wirral and West Lancashire revealed eight different haplogroups out of 13 tested for (Figure 7-1).

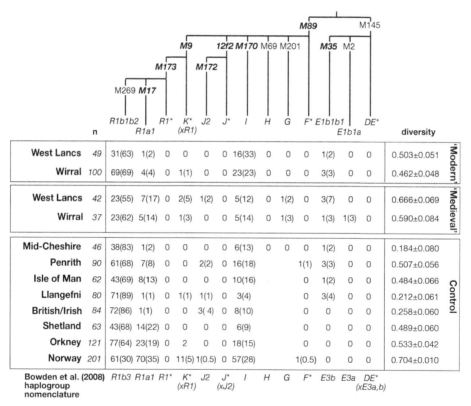

	n	R1b1b2	R1a1	R1*	K* (xR1)	J2	J*	I	H	G	F*	E1b1b	E1b1a	DE*	diversity	
West Lancs	49	31(63)	1(2)	0	0	0	0	16(33)	0	0	0	1(2)	0	0	0.503±0.051	'Modern'
Wirral	100	69(69)	4(4)	0	1(1)	0	0	23(23)	0	0	0	3(3)	0	0	0.462±0.048	
West Lancs	42	23(55)	7(17)	0	2(5)	1(2)	0	5(12)	0	1(2)	0	3(7)	0	0	0.666±0.069	'Medieval'
Wirral	37	23(62)	5(14)	0	1(3)	0	0	5(14)	0	1(3)	0	1(3)	1(3)	0	0.590±0.084	
Mid-Cheshire	46	38(83)	1(2)	0	0	0	0	6(13)	0	0	0	1(2)	0	0	0.184±0.080	Control
Penrith	90	61(68)	7(8)	0	0	2(2)	0	16(18)			1(1)	3(3)	0	0	0.507±0.056	
Isle of Man	62	43(69)	8(13)	0	0	0	0	10(16)			0	1(2)	0	0	0.484±0.066	
Llangefni	80	71(89)	1(1)	0	1(1)	1(1)	0	3(4)			0	3(4)	0	0	0.212±0.061	
British/Irish	84	72(86)	1(1)	0	0	3(4)	0	8(10)			0	0	0	0	0.258±0.060	
Shetland	63	43(68)	14(22)	0	0	0	0	6(9)			0	0	0	0	0.489±0.060	
Orkney	121	77(64)	23(19)	0	2	0	0	18(15)			0	0	0	0	0.533±0.042	
Norway	201	61(30)	70(35)	0	11(5)	1(0.5)	0	57(28)			1(0.5)	0	0	0	0.704±0.010	
Bowden et al. (2008) haplogroup nomenclature		R1b3	R1a1	R1*	K* (xR1)	J2	J* (xJ2)	I	H	G	F*	E3b	E3a	DE* (xE3a,b)		

FIGURE 7-1. *Haplogroup frequencies in the different populations. At the top is a tree of of Y chromosome haplogroups, and below are given the number of Y-chromosomes carrying each haplogroup, with percentages in parentheses. n: population sample size. A measure of genetic diversity ranging between 0 and 1 is also given. Two alternative haplogroup naming systems are included in the figure. Those on the tree are the current names, while those at the bottom are those used in our Molecular Biology and Evolution paper published in 2008*[106].

106. Bowden, G.R., Balaresque, P., King, T.E., Hansen, Z., Lee, A.C., Pergl-Wilson, G., Hurley, E., Roberts, S.J., Waite, P., Jesch, J., Jones, A.L.,Thomas, M.G., Harding, S.E. and Jobling, M.A. (2008) Excavating past population structures by surname-based sampling: the genetic legacy of the Vikings in North West England. *Molecular Biology and Evolution* volume 25, pages 301-309.

These haplogroups observed for Wirral and West Lancashire are typical of western European Y chromosomes, with one exception: a single example of a haplogroup known as E1b1a (previously known as E3a) in the Wirral 'medieval' sample. This haplogroup is typical of sub-Saharan African populations and in this case probably represents an African migrant, but not a recent one, since the man carrying the chromosome reports at least four generations of residence in the Wirral. It may, like a haplogroup A1a chromosome found previously in a man with ancestry in Yorkshire[107], represent African presence via the Roman occupation, or the Atlantic slave trade, and the proximity of the Wirral to the port of Liverpool might suggest the latter.

Comparison of the haplogroup profiles for 'modern' Wirral and West Lancashire with the other test sites are shown in Figure 7-2, but what is most remarkable is the difference between the medieval samples shown where there is not only an increase in gene diversity of the medieval samples with a greater number of haplogroups, but the proportion of haplogroup R1a1, common in Norway but relatively rare in areas of the British Isles not settled by people of Norwegian origin shows a marked increase to a level of ~17% in West Lancashire and ~14% in Wirral (Figure 7-3).

107. King,T.E., Parkin, E.J., Swinfield, G., Cruciani, F., Scozzari, R., Rosa, A., Lim, S.K., Xue, Y., Tyler-Smith, C., Jobling, M.A. (2007) Africans in Yorkshire? The deepest-rooting clade of the Y phylogeny within an English genealogy. *European Journal of Human Genetics*, volume 15, pages 288-293.

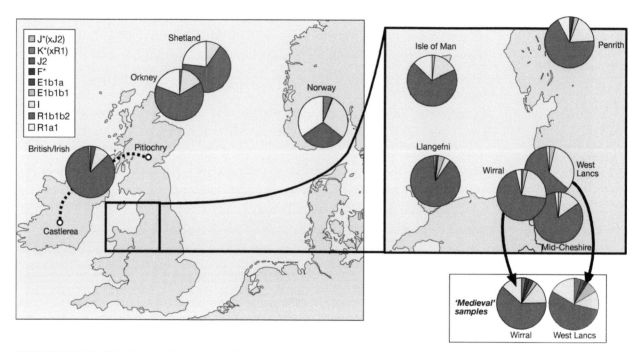

FIGURE 7-2. *Haplogroup distributions. Modern and medieval samples for Wirral and West Lancashire have significantly different sets of Y-chromosome haplogroups: for example, the proportion of a group known as 'R1a1', common in Norway, is significantly higher.*

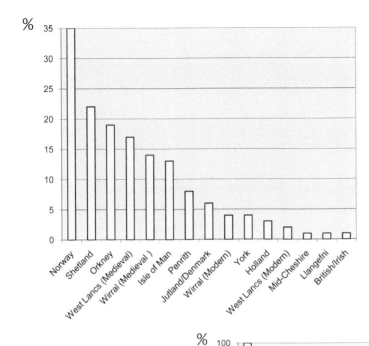

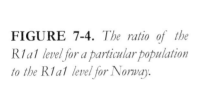

FIGURE 7-3. *Distributions of haplogroup R1a1. Medieval samples for Wirral and West Lancashire show high levels of haplogroup R1a1 (14% and 17% respectively), comparable to those levels in modern Orkney and Shetland[108].*

FIGURE 7-4. *The ratio of the R1a1 level for a particular population to the R1a1 level for Norway.*

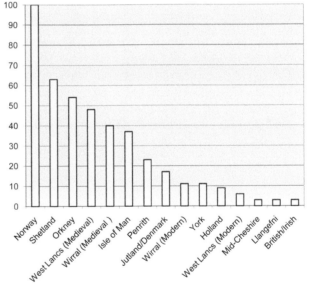

108. Capelli, C., Redhead, N., Abernethy, J.K., Gratrix, F., Wilson, J.F., Moen, T., Hervig, T., Richards, M., Stumpf, M.P.H., Underhill, P.A., Bradshaw, P., Shaha, A., Thomas, M.G., Bradman, N. and Goldstein, D.B. (2003) A Y-Chromosome census of the British Isles. *Current. Biology,* volume 13, pages 979-984.

These observations seem compatible with a greater Norse Viking contribution to the 'medieval' than to the 'modern' Wirral and West Lancashire samples.

Population differentiation tests

Statistical comparisons of genetic distances between populations based on the haplogroup distributions were then carried out between the modern and medieval Wirral and West Lancashire populations with Norway, the sample representing the putative population of the western British Isles prior to the Viking incursions (the Castlerea/Pitlochry sample), the Isle of Man, Llangefni, Penrith, Orkney and Shetland and shown in the plot of Figure 7-5. The same analysis was repeated but based on statistical comparisons of the haplotype data to give the graph of Figure 7-6. In both diagrams the 'medieval' Wirral and West Lancashire samples show a distinct shift towards Norway compared with their modern counterparts and lie closer to samples from the Isle of Man and Orkney, known to have experienced significant Scandinavian input.

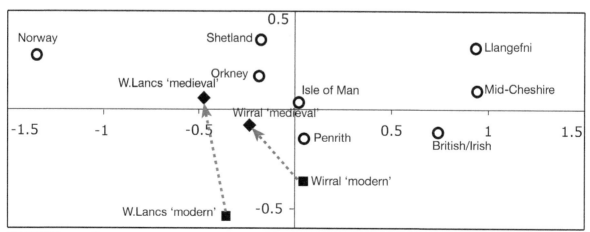

FIGURE 7-5. *Representation of genetic distances among populations, based on haplogroup distribution data.*

FIGURE 7-6. *Representation of genetic distances among populations, based on haplotype distribution data.*

Admixture analysis

Population admixture analysis was then performed on the haplotype and haplogroup data using the method of Helgason as described on pages 81-82 to estimate what proportions in the modern populations of Wirral and West Lancashire derive from a Norse source compared to a western British Isles source and how that changes for the 'medieval' samples.

The probability that a particular haplotype from the admixed population originated from the Norwegian parental population is determined by its relative frequency in Norwegian and British/Irish samples. If the haplotype is not present in the parental populations, this probability is derived from the relative frequency of the most closely related haplotypes in terms of mutational differences.

Where comparisons can be made, the proportions of Scandinavian admixture that we observe differ from those seen in a previous study[109]: for Shetland we observed 41% Scandinavian ancestry compared to the previously published figure of 44%, and the corresponding figures for Orkney are 50% compared to 31%. Such differences may reflect differences in the compositions of the parental and hybrid samples, and in the marker resolution – we used greater numbers of both binary markers and STRs. The lowest proportions of Scandinavian admixture among our samples are seen in Llangefni and Mid-Cheshire, at 10% and 21% respectively, and a higher proportion is seen in the Isle of Man, where there is a known

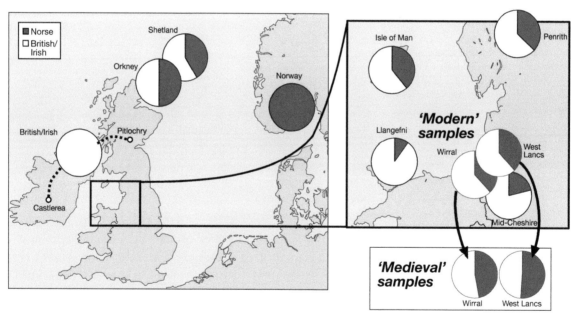

FIGURE 7-7. *Population admixture analysis of northwestern British Isles. Note that each admixture proportion is accompanied by a wide range of uncertainty, a fact not captured in this pie-chart representation.*

109. Goodacre S., Helgason A., Nicholson, J., Southam, L., Ferguson, L., Hickey, E., Vega, E., Stefánsson, K., Ward, R. and Sykes, B. (2005) Genetic evidence for a family-based Scandinavian settlement of Shetland and Orkney during the Viking periods, *Heredity* volume 95, pages 129-135.

history of Viking presence. The 'modern' samples from Wirral and West Lancashire both show 38% Scandinavian admixture, significantly higher than the nearby sample from Mid-Cheshire. This is consistent with the historical and place-name evidence for greater Viking presence in Wirral and West Lancashire than in Mid-Cheshire. The 'medieval' samples from both Wirral and West Lancashire show significantly increased proportions - 47% and 51% respectively - of Scandinavian ancestry compared to the 38% seen in their 'modern' equivalents (Figure 7-7). It seems likely that this striking difference is due to the sampling of a set of lineages that more closely reflects past patterns than sampling on the basis of recent ancestry in the area.

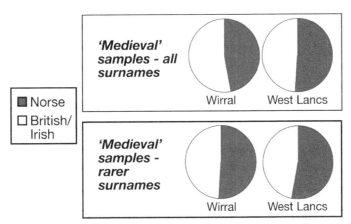

FIGURE 7-8. *Admixture results for rarer surnames. Men bearing names occurring more than 20,000 times in the British Isles have been excluded.*

Rarer surnames

The surnames that we used to ascertain the 'medieval' sample cover a wide range of frequencies - from *Otty*, with only 146 bearers in 1998 (www.spatial-literacy.org/UCLnames/) to *Brown*, with 242,765. The more frequent names are relatively widespread in Britain, and may provide less reliable links to medieval presence in the specific regions under study. To address this, we sub-sampled from the two 'medieval' samples by removing surnames with frequencies of greater than 20,000, resulting in reduced sample sizes of

26 and 30 for Wirral and West Lancashire respectively. We then repeated the admixture analysis. In both cases, the proportions of Scandinavian admixture increase significantly: from 47% to 51% in the Wirral, and from 51% to 53% in West Lancashire (Figure 7-8). This increase further validates the approach of surname-based ascertainment.

Anglo-Saxon/Danish Viking contribution to the admixture of North West England is relatively small

In the above admixture analyses we have made the "Western British isles" assumption of Helgason and Goodacre that the two parent populations are Norse and indigenous British and have neglected the possible contribution of DNA from Anglo-Saxon and Danish Viking settlers in the North West to the admixture.

As we have discussed already in Chapter 5 this may be a reasonably valid assumption as the work of Capelli and coworkers[110] has shown first of all that the haplogroup and haplotype distributions from modern Denmark and northern Germany are virtually impossible to distinguish from each other – as we have noted already they used a combined term of "invader" to describe the collective input from Anglo Saxon and Danish settlers. Secondly they also found that only in areas in England where there are significant numbers of place names of Scandinavian origin - such as Yorkshire, East Anglia and East Midlands - is this "invader" signal significant, implying that the Anglo-Saxon contribution to the gene pool of England - particularly western England - is relatively small. The high levels of R1a1 in the northwest of England are commensurate with Norse Viking rather than Anglo Saxon or even Danish Viking invaders.

---

110. Capelli, C., Redhead, N., Abernethy, J.K., Gratrix, F., Wilson, J.F., Moen, T., Hervig, T., Richards, M., Stumpf, M.P.H., Underhill, P.A., Bradshaw, P., Shaha, A., Thomas, M.G., Bradman, N. and Goldstein, D.B. (2003) A Y-Chromosome census of the British Isles. *Current Biology*, volume 13, pages 979-984.

Conclusion: population genetic analyses

The results strongly indicate that the northwest of England was once heavily populated by Scandinavian settlers, reinforcing the other evidence from place-names, archaeology and historical records. The results also show that Wirral and West Lancashire are similar to each other both for modern and medieval samples and are significantly different from North Wales and even central Cheshire. Over the centuries the River Mersey clearly has not been a barrier to significant population movement or exchange, contrasting with the River Dee.

Chapter 8
INDIVIDUAL RESULTS AND PUBLIC ENGAGEMENT

An important part of the project was providing a strong interface with the public and this was brought about by many public lectures reinforced by radio and TV broadcasts and newspaper articles. The public lectures included presentations at Hurlston Hall, Ormskirk in November 2002 and alsoNovember 2003, at Wirral Grammar School in March 2003, at the Athenaeum, Liverpool in November 2003 and to the Merseyside Young Archaeologists group in January 2002 and 2003.

Through these lectures and media contacts, reinforced by individual correspondence with the public, many people have become aware of the scientific basis behind DNA, chromosomes, genetics and inheritance, and the links with ancestry, history, linguistics and archaeology.

The culmination of the research project was the dissemination of the finds in a presentation at Knowsley, West Lancashire November 27th 2007 and subsequent publication in the high impact scientific journal *Molecular Biology and Evolution* published by Oxford University Press in February 2008. The host for the event was Patrick Waite, Chairman of the West Lancashire Heritage Association, and it was attended by nearly 200 guests including the Mayors of Wirral and the Chair of West Lancashire District Council (Figure 8-1).

The meeting started with an introduction by Mr. Waite followed by a description of the background of the project by Professor Steve Harding, the results by Professor Mark Jobling and a description of what Norse life might have been like in the northwest of England by Professor Judith Jesch. This was accompanied by displays given by re-enactment expert Kevin Taylor from Scarisbrick near Southport. It was followed in February 2008 by a presentation at St. Bridget's Church, West Kirby in which nearly 300 members of the public attended, raising nearly £3000 towards restoration work for the church[111].

111. http://www.nottingham.ac.uk/~sczsteve/St_Bridgets.txt

(a)

(b)

(c)

(d)

FIGURE 8-1. *Presentation at Knowsley Hall: (a) Introduction by the Chairman of the West Lancashire Heritage Association, Patrick Waite (b) Professor Steve Harding presents the background behind the survey (c) Professor Mark Jobling gives the results (d) Professor Judith Jesch talks about what life may have been like in the Scandinavian northwest of England.*

The results from the research were reported in the media and an article was subsequently published with Archaeologist Dr. David Griffiths to put the results in the context of the wealth of Viking age Archaeology in the northwest of England[112].

All people taking part in the survey were given a copy of their Y-chromosome data, a map of Europe showing where they had matches for their Y-chromosome haplotype, and an explanatory letter and a sheet showing where their top match frequencies were. This caused a great deal of interest even though it became clear to most that it is impossible to be definitive whether an individual is descended from a Viking or not, but we highlight three interesting examples here.

Tony Tottey, Brian Totty and Peter Forshaw

First of all we caught up with two Wirral men of similar surname - Brian Totty from Heswall and Neston and Tony Tottey from Moreton who both had identical haplotypes showing strongest match frequencies across Scandinavia[113] (Figure 8-2). Their results were particularly intriguing because of the similarity of their surname and because Tony's paternal uncle, a Mr. Gordon Tottey from West Kirby - now long since deceased - had featured in a newspaper article some 30 years previously about what he thought were his Norwegian roots (Figure 8-3). Gordon would have had the same Y-haplogroup and haplotype as Tony.

When asked about what he thought about the matches for his own Y-haplotype Tony said the following :

'I was not too surprised because there was an article a few years ago in the local paper about how people had suggested that my Uncle Gordon could have had links to the Vikings. He was actually a very laid back type of person.

---

112. Griffiths, D., Harding, S.E. and Jobling, M.A. (2008) Looking for Vikings in North West England. *British Archaeology* volume 103, pages 18-25.

113. It is possible to see the interview with Brian and Totty on this link: http://www.wirral-mbc.gov.uk/Vikings/StevesVids/22%20DNA.wmv as part of the Wirral Learning Grid programme for schools.

In his early days he worked as a shipwright in Cammell Lairds Shipyard – apparently on one occasion whilst working on Norwegian shipping this Norwegian chap came up to him and said "I could be your cousin". He looked at this man and apparently it was like looking in a mirror - and lo and behold it turned out his name was also Tottey, or a Norwegian form of this.'

'I think Uncle Gordon was taken aback because people had said he could be related to the Vikings. He always played down the idea but that was him - he took everything as it came. Nothing phased him. He was quite an easy going sort of person and I think, quite honestly, he would have just said *OK that proves it I could be related to the Vikings*. He was that type of person.'

FIGURE 8-2. *Headline from an article from Monday March 15th, 1971 in the Liverpool Daily Post about Gordon Tottey*[114].

Red Rocks, West Kirby, Caldy, Frankby, Thurstaston.

The last of the Wirral Vikings...

by Jim Barrow
Pictures: Eddie Barford

THE NORSEMAN
Mr Gordon Tottey, his forefathers landed with Vikings invaders.

RED ROCKS is a tiny finger of sandstone pointing out into the Irish Sea from the extreme western point of the Wirral Peninsula.

Bright sunshine warmed the soft redrock and glinted on the tiny waves breaking on them, as Eddie Barford, and I started our walk—with more than 100 miles of Cheshire and North Wales ahead of us.

Red Rocks, our starting point, looked east to Hoylake, and a deserted sweep of beach. To the West on Hilbre Island, in the Dee Estuary, now a bird sanctuary, the building sparkled white in sunshine.

In front of us we saw an elderly man, his trousers tucked into wellington boots, and well muffled against the wind swinging a golf club, and driving a ball hard down the sand before marching after it.

With difficulty we managed to catch up with our beach golfer, who turned out to be 72-year-old Mr Gordon Tottey, whose family are known to have lived in West Kirby at the time the Domesday Book was compiled.

He told us: "I've been told I'm descended from the Vikings, who went to Ireland, and then came over here to West Kirby, but I don't see how anyone can know."

Although Mr Tottey scoffs at the possible link with the Vikings, he does recall the shock he got while working on a Norwegian ship. He says: "One of the men on board, who was a Norwegian, came up to me and said: 'I might be your cousin'. I looked at his face and I might have been looking into a mirror—his name was Tottey."

Mr Tottey was born in a black and white timbered cottage on the site of the present railway station, and for more than 30 years he ran a sweets and tobacconist's shop.

He pointed to the promenade and told us: "One of the houses was used by a man called Mr Mines who used it to build a very frail looking aircraft to compete in the Manchester Air Race of 1908."

Mr Tottey still fishes the Dee, but says that there aren't half as many fish as there used to be and blames the pollution in the Dee and the seals which get into the estuary.

FIGURE 8-3. *Tony Tottey (left) and Brian Totty. Still from a recorded interview on "Viking DNA"*[115] *for the Vikings in Wirral schools website, courtesy of The Wirral Learning Grid, http://www.wirral-mbc.gov.uk*

114. Barrow, J. (1971) The last of the Wirral Vikings, *Daily Post*, 15/3/1971, page 5. Link: https://www.nottingham.ac.uk/~sczsteve/Post_15Mar71.jpg
115. http://www.wirral-mbc.gov.uk/Vikings/StevesVids/22%20DNA.wmv

Then we asked Brian Totty if he was surprised about his results, and he replied as follows:

'Well, I wasn't surprised, but rather delighted to get the proof of something which had been suggested to me for many years by a local historian. His name was Canon Lee who was the rector of Heswall Church and he would be poring over the Parish Registers and whenever I bumped into him he would say *Here comes Totty the Viking*. I used to laugh and think he was winding me up. I accepted in good faith what he said about the Parish Register and that goes back five centuries or so to the 1650's but these results now take us back another seven hundred years would it be? This has scientifically proven what the suggestion was so the result has proved something that couldn't be established with certainty simply through anecdotes in Parish Registers'.

The surprising thing about both Tony and Brian's reaction to their results - was they were not surprised!

My Viking Dad with my Viking dog!

Viking beer!

FIGURE 8-4.
Peter Forshaw from Irby, featured as "My Viking Dad" in a school project by his 10 year old daughter Abigail.

Another participant who came up with his strongest frequency matches for his Y-chromosome haplotype (within haplogroup I1) across Scandinavia was Peter Forshaw from Irby, Wirral….Peter's daughter Abigail was so intrigued by her father's result it inspired her to write a school project about him: "*My Viking Dad, his Viking Dog and his Viking Beer*" (Figure 8-4).

Public Lectures

The team also engaged in lectures to community groups, schools, local societies, and presentations were given both during and after the end of the project as follows:

Merseyside Young Archaeologists, Maritime Museum Liverpool (26/1/2002)

Hoylake, Inner Wheel Inter Club, Hoylake (12/3/2002)

Birkenhead History Society, Williamson Museum (18/4/2002)

Hoylake and District Civic Society (11/6/2002)

Wirral Grammar School (11/7/2002)

Grosvenor Museum, Chester (10/8/2002)

Heswall & District Rotary Club (19/9/2002)

Bromborough Society (27/9/2002)

West Lancashire Heritage Association, Hurlston Hall Golf Club, Ormskirk (13/11/2002)

Mayer Trust Lecture, Bebington (11/12/2002)

Merseyside Young Archaeologists, Runcorn Priory (25/1/2003)

School of Archaeology, University of Nottingham (26/2/2003)

Names, Genetics - an Interdisciplinary Workshop, University of Nottingham (26/3/2002)

Wallasey Historical Society (13/3/2003)

Friends of Storeton Woods (18/3/2003)

Liverpool Athenaeum Club (19/11/2003)

Ormskirk Advertiser Lecture, Hurlston Hall (19/11/2003)

Neston Historical Society (11/12/2003)

Formby Civic Society (23/1/2004)

Thurstaston Visitor Centre (26/4/2004)

Poulton Hall, Bromborough (4/7/2004)

University of Newcastle, Conference on DNA and the Middle Ages (22/10/2004)

Merseyside Archaeological Society (20/1/2005)

Friends of Hoylake and Meols Gardens and Open Spaces (2/4/2005)

Midlands Viking Symposium, University of Nottingham (9/4/2005)

West Wirral Rotary Club Charter Anniversary Dinner (12/10/2005)

Wirral Schools Learning Grid Launch, Tranmere Rovers Conference Room (12/12/2006)

University of Chester History Society (27/2/2007)

University of Leicester Midland Viking Symposium (28/4/2007)

North Wirral Rotary Club (2/10/2007)

University of Iceland at Reykjavik (23/4/2008)

Irby, Thurstaston and Pensby Amenity Society (4/6/2008)

University of Stavanger, Norway (28/9/2008)

Family History Society of Cheshire, West Kirby (12/11/2008)

NICE08 (Nordic Cultural Events Festival, Liverpool) Schools Workshops - Bishop Wilson School, Neston and Lady Immaculate School, Everton Valley (25/11/2008)

LINC (Lectures in the Nordic Community) Lecture, Scandinavian Church, Liverpool (25/11/2008)

NICE08 Festival Public DNA testing, Scandinavian Church, Liverpool (26/11/2008) – Figure 8-6

Chester Archaeological Society, Grosvenor Museum (4/2/2009)

Archaeology Festival, Cardiff (9/2/2009)

North Wirral Probus Club, Hoylake (17/3/2009)

Wirral Bookfest Talk, Wallasey Library (12/10/2009)

British Archaeology North West Regional Group Conference, Lytham-St. Annes (24/10/2009)

University of Oslo (10/6/2010)

Radium Sykkehus, Oslo (11/6/2010)

Archaeological Festival day, Bidston (17/7/2010)

Wirral Bookfest Talk, Bebington Library (14/10/2010)

Vinland Congress, Chicago (16/10/2010)

Chester Viking Conference (20/11/2010)

The Wirral News ● Wednesday, September 16, 2009   ● **Visit our website:** www.wirralnews.co.uk   HES

More links to our Viking past

MORE Vikings have been found in Wirral!

Scientists found two men from Meols shared identical historical links to Scandinavia during DNA testing.

Bizarrely one – Stan Royden – is married to a Norwegian woman, Mette, and is chairman of the committee for the Scandinavian Church in Liverpool.

The second – Roy Shuttleworth – is secretary of the Friends of Meols Park and has been looking into the area's Viking past.

Although not related, the men were found to have very similar chromosome types.

Their strongest DNA link was to Gotland, an island off the east coast of Sweden.

The findings were released this week and are the result of a DNA ancestry event held last November as part of the Nordic Festival in Liverpool.

Viking expert Professor Steve Harding from Nottingham University and colleagues Professor Mark Jobling and Dr Turi King at Leicester University tested 195 men free of charge.

A genetic survey carried out last year by scientists from Nottingham and Leicester Universities and University College London found that up to 50% of the DNA of men from old Wirral families was Norse in origin.

Mr Royden, 64, said: "I was quite surprised because I thought I was all Anglo-Saxon!

"I've always felt an affinity with the place – my first job was in Norway, which is where I met my wife Mette."

In another bizarre coincidence, the Scandinavian Church Stan is involved with is in the Diocese of Gotland and comes under the jur-

By LORNA HUGHES

isdiction of the Bishop of Gotland.

Mr Shuttleworth, 56, said he was "totally amazed" by the results.

He said: "I couldn't believe it , especially as we're working on a Viking project in Meols."

Only men can be tested for the Viking link because the test is based on DNA from the male Y chromosome which is passed down the paternal line from generation to generation, with little or no change.

Professor Harding said: "The results for Stan and Roy showed the match in both cases was Gotland in Sweden, where 15% of men have the same Y chromosome type as Stan and Roy.

"They also have matches elsewhere around Scandinavia.

"We can't say for sure but there is a very good chance they are both carrying the Y-chromosome of a Viking. Don't forget the Vikings moved all a lot, so they didn't necessarily come from just one place.

"They're just ordinary men but in every cell they have this link to Scandinavia."

●A research paper on the Wirral genetic survey was published last year. Find out more at http://www.nottingham.ac.uk/~sczsteve .

Professor Harding is giving a talk on Viking Wirral and Viking genes at Wallasey Central Library between 2.30 and 4pm on Mon, Oct 12.

For more information call 639 2334.

● Roy Shuttleworth (left) and Stan Royden (right) with Prof Stephen Harding.

FIGURE 8-6. *A special event was held on 26th November 2008 at the Scandinavian Church, Park Lane Liverpool as part of the Nordic Cultural Events Festival. Free Y-chromosome testing was offered to male participants of the Festival, and yielded some interesting results, such as for Stan Royden and Roy Shuttleworth from Meols. Stan, and his Norwegian wife Mette help run the Church and Roy, Secretary of the Friends of Meols Park Organisation, is a keen promoter of the areas Viking Heritage. Both have similar Y-haplotypes within haplogroup I1 with their strongest matches in Gotland, Sweden[116].*

---

116. Article reprinted from *Wirral News* courtesy of Trinity Mirror Copyright.

Chapter 9
CONCLUSIONS AND PERSPECTIVES

Region heavily settled by Norse Viking settlers

In this study, we have focused on a region in which historical and place-names evidence suggests a past presence of Viking incomers and demonstrated that samples ascertained using surnames present in medieval times show a markedly greater proportion of Scandinavian ancestry. Our genetic analysis appears to confirm a belief, based on archaeological, place-name and historical evidence, that the northwest of England was once heavily settled by Norse Vikings, many of whom were refugees expelled from Ireland in AD 902[117]. Samples from "Old Wirral" and "Old West Lancashire" revealed up to ~50% Norse admixture in the Y chromosomes of men from old families in Wirral and old families in West Lancashire.

Prior to arriving in Wirral, Ingimund's Scandinavian refugees expelled from Ireland are reported to have first made an attempt to settle in Anglesey: recent archaeological evidence for some Scandinavian activity[118] supports this, but we observe a low level of Scandinavian admixture there. In contrast, significant and lasting settlements appear to have been made in the northwest of England and the Isle of Man with a strong contribution to the genetic admixture. The detailed relationship of the settlements described in our study to other Viking activities in the Irish Sea region still remains to be unravelled, but the combination of linguistic, archaeological, historical, literary, and, where possible, genetic evidence will continue to provide valuable clues.

Maternal DNA contributions

Apart from the Y-chromosome, passed from father to son with little change, we described an analogous maternally inherited piece of DNA,

117. Wainwright, F.T. (1975) *Scandinavian England: Collected Papers*. Ed. Finberg, H.P.R., Phillimore, Chichester, UK.
118. Redknap, M. (2000) *Vikings in Wales*. National Museums and Galleries of Wales, Cardiff.

mitochondrial DNA, which passes from mother to children, and this too can be used to compare population movements. As we discussed in Chapter 5, in a fascinating study recently published by Helgason and coworkers[119] on Iceland it was shown that whereas the male population showed a strong correlation with Norway the female population had a high affinity with the Gaelic countries leading to much interesting discussion as to why that should be so. This was mirrored in a similar study on the Orkneys[120]. This could also prove a very fruitful area of research for the northwest of England, although it would have to be done without the use of surnames as a guide.

Surname-based sampling may help in other regions

In our study focusing on two particular areas of North West England, the association of Y chromosomes with patrilineal surnames, which were established in England some 700 years ago[121], has been shown to provide a powerful tool to investigate changes in population structure over the last few centuries[122] - an era in which population growth and migration have been considerable. Our findings suggest that the approach of surname-based ascertainment is promising and may provide a means to address the issue of population change through time, as well as some specific questions in the histories of regions having patrilineal surnames.

119. Helgason, A., Sigurdardóttir, S., Gulcher, J.R., Ward, R. and Stefansson, K. (2000) mtDNA and the origin of the Icelanders: deciphering signals of recent population history. *American Journal of Human Genetics* volume 66, pages 999-1016; Helgason, A., Sigurdardóttir, S., Nicholson, J., Sykes, B., Hill, E.W., Bradley, D.G., Bosnes,V., Gulcher, J.R., Ward, R. and Stefánsson, K. (2000) Estimating Scandinavian and Gaelic ancestry in the male settlers of Iceland. *American Journal of Human Genetics* volume 67, pages 697–717.
120. Goodacre S., Helgason A., Nicholson, J., Southam, L., Ferguson, L., Hickey, E., Vega, E., Stefánsson, K., Ward, R. and Sykes, B. (2005) Genetic evidence for a family-based Scandinavian settlement of Shetland and Orkney during the Viking periods. *Heredity*, volume 95, pages 129-135.
121. McKinley, R. (1981) *The Surnames of Lancashire*. Surnames series volume 4, Leopard's Head Press, London.
122. Manni, F., Toupance, B., Sabbagh, A. and Heyer, E. (2005) New method for surname studies of ancient patrilineal population structures, and possible application to improvement of Y-chromosome sampling. *American Journal of Physical Anthropology*, volume 126, pages 214–228.

The robustness of conclusions will be improved by the inclusion of both modern and medieval samples from places not thought to have experienced particular population incursions, such as Mid-Cheshire in our study (surname-based sampling from this region will form part of a future project). The approach relies on the existence of appropriate historical surname lists, which are plentiful, and, as we have shown, can also be supplemented by lists based on place-names. Analysis of the distributions of surnames across the British Isles will also prove valuable in the future selection of

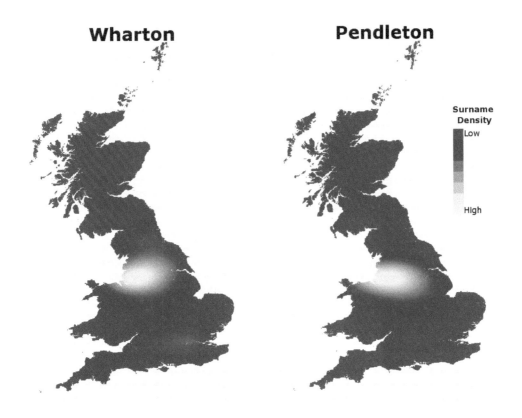

FIGURE 9-1. *Maps showing the surname "hot-spots" for Wharton (left) and Pendleton (right) in 1881. Both surnames are clearly highly concentrated in West Lancashire and Wirral. Such maps are used to discern the core areas of a surname's distribution and can inform future volunteer recruitment*[123].

123. Reproduced courtesy of James Cheshire, University College, London.

volunteers. For example the University College London in conjunction with the National Trust now have a facility for assessing distributions of surname frequencies in the British Isles[124] - one team at UCL have developed the new method of *Surname Cores*[125] and two examples are given in Figure 9-1 for surnames with "hot-spots" in Wirral and West Lancashire. However, regions where the time depth of patrilineal surnames is shallow, such as Wales, where patronymic systems have persisted to some extent until the 19th Century[126], will be more difficult to study in this way.

To maximize the benefits of surname-based ascertainment, care needs to be taken in sampling. The effects on the Scandinavian admixture proportions of subsampling from our medieval samples suggest that common surnames should be avoided where possible. In this study, we sampled one male per surname for each of the two regions studied. However, previous studies of the relationships between surnames and Y haplotypes have shown the effects of non-paternities in breaking the link between the two[127], and a recent non-paternity could introduce the Y haplotype of an outsider into a surname which has medieval roots in a study area. To obviate this problem, it may be desirable to sample more than one male per surname and to treat the majority haplotype as a founder. In fact, the chance of sampling recently introgressed Y chromosomes increases as sample size per surname increases but the power to distinguish such chromosomes from the earlier founding types also increases because a consensus Y chromosome type can more readily be derived.

The use of patrilineal surnames in recruitment ascertainment provides both

124. http://gbnames.publicprofiler.org/

125. Cheshire, J.A., Mateos, P. and Longley, P.A. (2009) Family names as indicators of Britain's changing regional geography, *UCL Working Papers Series*, Paper 149, pages 1-63, ISSN 1467-1298.

126. McKinley, R.A. (1990) *A History of British Surnames*. Longman Press, London.

127. Sykes, B. and Irven, C. (2000) Surnames and the Y chromosome. *American Journal of Human Genetics* volume 66, pages 1417–1419; King, T.E., Ballereau, S.J., Schürer, K. and Jobling, M.A. (2006) Genetic signatures of coancestry within surnames. *Current Biology* volume 16, pages 384–388; McEvoy, B. and Bradley, D.G. (2006) Y-chromosomes and the extent of patrilineal ancestry in Irish surnames. *Human Genetics*, volume 119, pages 212–219.

the power of our approach and its major limitation because our analysis of the structure and history of thepopulations we have studied has been based on a single genetic locus, the Y chromosome. Nonetheless, in the absence of a time machine, the link between a haplotype and a cultural marker, the surname, may provide the only practical means for accessing the genetic composition of populations in the past. As well as allowing us to investigate the influence of migration and drift over the last few centuries in changing the population structure of Britain, the method should be applicable to other regions where surnames are patrilineal and suitable historical records survive.

Indeed, thanks to the generous support from the Wellcome Trust, the successful study of Wirral and West Lancashire is now being extended to North Lancashire, Cumbria and North Yorkshire in an attempt to see how far into medieval northern England the Norse settlers from the Irish Sea penetrated (Figure 9-2). Part of the study is focused on obtaining better control data from Scandinavia – looking for regional variations in Y-chromosome distributions within Norway, Sweden and Denmark and in particular what the old or Medieval profiles in these countries may have looked like - and a team has been set up to undertake this, including Dr. Berit Dupuy from the University of Oslo and two men with a strong interest in the history and heritage of Norway - businessmen Sigurd Aase from Haugesund and Harald Løvvik from Oslo. Dr. Dupuy and colleagues have already shown significant regional variations across Modern Norway[128]. Although patrilineal surnames are not so useful for volunteer recruitment in Scandinavia as they are relatively new there (and do not exist at all in Iceland) it has been possible with the help of local historical societies to recruit volunteers whose paternal ancestry goes back many generations in particular areas (Figure 9-3). The following areas in Norway for example have been identified: four western fjord areas (Sognefjorden, Hardangerfjorden, Ryfylkefjorden, More); three western city areas (Trondheim, Bergen, Stavanger) and three eastern areas (Hedemark, Namdalen, Gudbrandsdalen). The results for this will be reported in a future publication.

---

128. Dupuy B.M., Stenersen, M., Lu, T.T, and Olaisen, B. (2006) Geographical heterogeneity of Y-chromosomal lineages in Norway. *Forensic Science International*, volume 164, pages 10-19.

FIGURE 9-2. *Finding Vikings in North Lancashire, Cumbria and North Yorkshire. A new project funded by the Wellcome Trust explores how far Norse influence from the North West penetrated into Eastern England.*

Jakter på viking-gen

FIGURE 9-3. *Testing old Norway – headline from the Haugesunds Avis & Karmøy Lokal newspaper, 10th June 2010, and reprinted courtesy of Terje Størksen*[129].

Forskertrioen Stephen Harding (t.v.), Turi King og Mark Jobling trenger hjelp av norske menn til DNA-undersøkelsen. Den kan fint gjøn nomføres hjemme. En prøve av munnepiteceller fra innsiden av kinnet med denne pinnen er det som skal til. (Foto: Terje Størksen)

---

129. Terje.Storksen@h-avis.no

Bibliography

Adams, S.M., King, T.E., Bosch, E. and Jobling, M.A. (2006) The case of the unreliable SNP: recurrent back-mutation of Y-chromosomal marker P25 through gene conversion. *Forensic Science International*, volume 159, pages 14–20.

Bean, S.C. (2000) Silver ingot from Ness, Wirral. In: Cavill, P., Harding, S.E. and Jesch, J. (eds.) *Wirral and its Viking Heritage*. English Place-Name Society, Nottingham, UK, pages 17-18.

Bailey, R., Whalley, J., Bowden, A. and Tresise, G. (2006) A miniature Viking-age hogback from the Wirral. *Antiquarian Journal*, volume 86, pages 345-356.

Bennett, J.H.E. and Dewhurst, J.C. (1940) Quarter sessions records with other records of the Justices of the Peace for the County Palatine of Chester 1559-1760 together with a few earlier miscellaneous records deposited with the Cheshire County Council. *Records Society of Lancashire and Cheshire*, volume 94, pages 37-39.

Blackburn, M.A.S. (2006) Currency under the Vikings. Part 2 : The two Scandinavian kingdoms of the Danelaw, c. 895-964. *British Numismatic Journal*, volume 76, pages 204-226.

Booth, P. (1983) Calendar of Cheshire Trailbaston Proceedings 1353. *Cheshire History* 12, pages 24-28.

Bosch, E., Calafell, F., González-Neira, A., Flaiz, C., Mateu, E., Scheil, H.-G., Huckenbeck, W., Efremovska, L., Mikerezi, I., Xirotiris, N., Grasa, C., Schmidt, H. and Comas, D. (2006) Male and female lineages in the Balkans show a homogeneous landscape over linguistic barriers, except for the isolated Aromuns. *Annals of Human Genetics* volume 70, pages 459-487.

Bowden, G.R., Balaresque, P., King, T.E., Hansen, Z., Lee,A.C., Pergl-Wilson, G.,Hurley, E., Roberts, S.J., Waite, P., Jesch, J., Jones, A.L.,Thomas, M.G., Harding, S.E. and Jobling, M.A. (2008) Excavating past population structures by surname-based sampling: the genetic legacy of the Vikings in North West England. *Molecular Biology and Evolution* volume 25, pages 301-309.

Bu'Lock, J.D. (1958) Pre-Norman crosses of West Cheshire and the Norse settlements around the Irish Sea. *Transactions of the Lancashire and Cheshire Antiquarian Society*, volume 68, pages 1-11.

Capelli, C., Redhead, N., Abernethy, J.K., et al, (15 co-authors) (2003) A Y chromosome census of the British Isles. *Current Biology* volume 13, pages 979-984.

Cavill, P., Harding, S.E.and Jesch, J. (2000) *Wirral and its Viking Heritage*. English Place-Name Society, Nottingham, UK.

Cavill, P., Harding, S.E. and Jesch, J. (2003–2004), Revisiting Dingesmere, *Journal of the English Place Name Society*, volume 36, pages 25-38.

Chaix, R., Quintana-Murci, L., Hegay, T., Hammer, M.F., Mobasher, Z., Austerlitz, F. and Heyer, E. (2007) From social to genetic structures in central Asia. *Current Biology*, volume, pages 43-48.

Collingwood, W.G. (1928) Early monuments of West Kirby, In: Brownbill, J.ed. *West Kirby and Hilbre. A Parochial History*. Henry Young & Sons Ltd., Liverpool , UK, pages 14-26.

Dodgson, J.M. (1957) The background of Brunanburh, *Saga-book of the Viking Society*, volume 14, pages 303-316.

Dodgson, J.M. (1972) *The Place Names of Cheshire, Part Four*. Cambridge University Press, Cambridge, UK.

Dupuy B.M., Stenersen, M., Lu, T.T, and Olaisen, B. (2006) Geographical heterogeneity of Y-chromosomal lineages in Norway. *Forensic Science International*, volume164, pages 10-19.

Ekwall, E. (1922) *The Place-Names of Lancashire*. Manchester University Press, Manchester, UK.

Fellows-Jensen, G. (1992) Scandinavian place-names of the Irish Sea Province. In: Graham-Campbell J.A., ed. *Viking Treasure from the North-west: the Cuerdale Hoard in its Context*. National Museums and Galleries on Merseyside Occasional Papers, Liverpool, UK, pages 31-42.

Fryer P. (1984) *Staying Power: the History of Black People in Britain*. Pluto Press, London, UK.

Gamlin, H. (1897) *Twixt Mersey and Dee*. Marples, Liverpool, UK.

Goodacre, S., Helgason, A., Nicholson, J., Southam, L., Ferguson, L., Hickey, E., Vega, E., Stefansson, K., Ward, R., Sykes, B. (2005) Genetic evidence for a family-based Scandinavian settlement of Shetland and Orkney during the Viking periods. *Heredity*, volume 95, pages 129–135.

Graham-Campbell, J.A. (1992) *Viking Treasure from the North-west: the Cuerdale Hoard in its Context*. National Museums and Galleries on Merseyside Occasional Papers, Liverpool, UK.

Griffiths, D. (2010) *Vikings of the Irish Sea*. History Press, Stroud, UK.

Griffiths, D., Philpott, R.A. and Egan, G. (2007) *Meols. The Archaeology of the North Wirral Coast*. Oxford University School of Archaeology, Oxford, UK, pages 58-76.

Griffiths, D., Harding, S.E. and Jobling, M.A. (2008) Looking for Vikings in Northwest England, *British Archaeology* Nov/Dec, pages 18-25.

Harding, S.E. (2000) Locations and legends. In Cavill, P., Harding, S.E. and Jesch, J. (eds.) *Wirral and its Viking Heritage*. English Place-Name Society, Nottingham, UK, pages 100-124.

Harding, S.E. (2002) *Viking Mersey: Scandinavian Wirral, West Lancashire and Chester*. Countyvise, Birkenhead, UK.

Harding, S.E. (2000) *Ingimund's Saga. Norwegian Wirral*, Countyvise, Birkenhead, UK. 2nd edition, 2006.

Harding, S.E. (2007) The Wirral carrs and holms, *Journal of the English Place-Name Society*, volume 39, pages 45-57.

Helgason, A., Hickey, E., Goodacre, S., Bosnes, V., Stefansson, K., Ward, R. and Sykes, B. (2001) mtDNA and the islands of the North Atlantic: estimating the proportions of Norse and Gaelic ancestry. *American Journal of Human Genetics*, volume 68, pages 723-737.

Helgason, A., Hickey, E., Goodacre, S., Vega, E., Bosnes, V., Stefánsson, K., Ward, R., Sykes, B. (2001) mtDNA and the islands of the North Atlantic: Estimating the proportions of Norse and Gaelic ancestry. *American Journal of Human Genetics*, volume 68, pages 723-737.

Helgason, A., Sigurdardóttir, S., Gulcher, J.R., Ward, R. and Stefansson, K. (2000) mtDNA and the origin of the Icelanders: deciphering signals of recent population history. *American Journal of Human Genetics* volume 66, pages 999-1016.

Helgason. A., Sigurdardóttir, S., Nicholson, J., Sykes, B., Hill, E.W., Bradley, D.G., Bosnes,V., Gulcher, J.R., Ward, R. and Stefánsson, K. (2000) Estimating Scandinavian and Gaelic ancestry in the male settlers of Iceland. *American Journal of Human Genetics*, volume 67, pages 697-717.

Irvine, W.F. and Sanders, F. (1894) *Wirral Notes and Queries, Being Gleanings Historical and Antiquarian*. Willmer Brothers, Birkenhead, UK.

Jesch, J. (1991) *Women in the Viking Age*. Boydell Press, Woodbridge (New Jersey), USA.

Jobling, M.A. (2001) In the name of the father: surnames and genetics. *Trends in Genetics*, volume 17, pages 353–357.

Jobling, M.A., Hurles, M.E. and Tyler-Smith, C. (2004) *Human Evolutionary Genetics: Origins, Peoples and Disease*. Garland Science, New York, USA.

Jobling, M.A. and Tyler-Smith, C. (2003) The human Y chromosome: an evolutionary marker comes of age. *Nature Reviews Genetics*, volume 4, pages 598-612.

Keyser-Tracqui, C., Crubézy, E. and Ludes, B. (2003) Nuclear and mitochondrial DNA analysis of a 2,000-year-old necropolis in the Egyin Gol Valley of Mongolia. *American Journal of Human Genetics*, volume 73, pages 247-260.

King, T.E., Ballereau, S.J., Schürer, K. and Jobling, M.A. (2006) Genetic signatures of coancestry within surnames. *Current Biology*, volume 16, pages 384-388.

King, T.E., Parkin, E.J., Swinfield, G., Cruciani, F., Scozzari, R., Rosa, A, Lim, S.K., Xue, Y., Tyler-Smith, C. and Jobling, M.A. (2007) Africans in Yorkshire? The deepest-rooting clade of the Y phylogeny within an English genealogy. *European Journal of Human Genetics*, volume 15, pages 288-293.

Magnusson, Magnus (2000) *The Vikings*, Tempus Publishing Ltd., Stroud, UK.

Manni, F., Toupance, B., Sabbagh, A. and Heyer, E. (2005) New method for surname studies of ancient patrilineal population structures, and possible application to improvement of Y-chromosome sampling. *American Journal of Physical Anthropology* volume 126, pages 214-228.

McEvoy, B. and Bradley, D.G. (2006) Y-chromosomes and the extent of patrilineal ancestry in Irish surnames. *Human Genetics*, volume 119, pages 212-219.

McEvoy, B., Brady, C., Moore, L.T. and Bradley, D.G. (2006) The scale and nature of Viking settlement in Ireland from Y-chromosome admixture analysis. *European Journal of Human Genetics*. volume 14, pages 1288-1294.

McKinley, R. A. (1981) *The Surnames of Lancashire. Surnames series volume 4.* Leopard's Head Press, London, UK.

McKinley, R.A. (1990) *A History of British surnames.* Longman Press, London, UK.

Peet, G. (1991) Inhabitants of Ormskirk, Scarisbrick with Hurlton, Bickerstaffe, Burscough with Marton, Westhead with Lathom and Skelmersdale who promised to contribute to the stipend of the priest of the altar of Our Lady at Ormskirk, 1366. *Ormskirk District Family Historian*, volume 1, pages 2-6.

Philpott, R.A. and Adams, M.H. (2009) *Irby, Wirral: Excavations on a Late Prehistoric, Romano-British and Medieval Site*, 1987-96, National Museums Liverpool, UK.

Reaney, P.H. (1927) *A Grammar of the Dialect of Penrith (Cumberland): Descriptive and Historical, with Specimens and Glossary.* Manchester University Press, Manchester, UK.

Redknap, M. (2000) *Vikings in Wales*. National Museums and Galleries of Wales, Cardiff, UK.

Richards, J.D. (2004) *Viking Age England*. Tempus, Stroud, UK.

Roberts, S.J. (2002) *A History of Wirral*. Phillimore, Chichester, UK.

Schneider, S., Roessli, D. and Excoffier, L. (2000) *Arlequin Version 2.0: a Software for Population Genetics Data Analysis.* Genetics and Biometry Laboratory, University of Geneva, Switzerland.

Sturges, C.M. and Haggett, B.C. (1987) *Inheritance of English surnames.* Hawgood Computing, London, UK.

Sulley, P. (1889) *The Hundred of Wirral.* Haram & Co., Birkenhead, UK, pages 212-216.

Sykes, B. and Irven, C. (2000) Surnames and the Y chromosome. *American Journal of Human Genetics*, volume 66, pages 1417-1419.

Thomas, M.G, Bradman, N., and Flinn, H.M. (2000) High throughput analysis of 10 microsatellite and 11 diallelic polymorphisms on the human Y chromosome. *Human Genetics*, volume 105, pages 577-581.

Thomas, M.G., Stumpf, M.P. and Harke, H. (2006) Evidence for an apartheid-like social structure in early Anglo-Saxon England. *Proceedings of the Royal Society: Biological Science*, volume 273, pages 2651-2657.

Topf, A.L., Gilbert, M.T., Dumbacher, J.P. and Hoelzel, A.R. (2006) Tracing the phylogeography of human populations in Britain based on 4th-11th Century mtDNA genotypes. *Molecular Biology and Evolution*, volume 23, pages 152-161.

Wainwright, F.T. (1942) North-west Mercia AD 871-924. *Transactions of the Historical Society of Lancashire and Cheshire*, volume 94, pages 3–55. Reprinted (in part) in Cavill et al. (2000), pages 19-42.

Wainwright, F.T. (1943) Wirral field names. *Antiquity*, volume 27, pages 57–66. Repeated (in part) in Cavill et al. (2000), pages 98-99.

Wainwright, F.T. (1945) The Scandinavians in Lancashire. *Transactions of the Lancashire and Cheshire Antiquarian Society*, volume 58, pages 71–116.

Wainwright, F.T. (1948) Ingimund's invasion. *English Historical Reviews*, volume 247, pages 145-167.

Wainwright, F.T. (1975) *Scandinavian England: collected papers*. Edited by Finberg H.P.R., Phillimore, Chichester, UK.

Wawn, A. (2000) Vikings and Victorians. In Cavill, P., Harding, S.E. and Jesch, *Wirral and its Viking Heritage*. English Place-Name Society, Nottingham, UK, pages 100-107.

White, R.H. (1986) Viking Period Sculpture at Neston, Chester, *Journal of the Chester Archaeological Society*, volume 69, pages 45-58.

Wilson, E. (1979) Sir Gawain and the Green Knight and the Stanley family of Stanley, Storeton, and Hooton. *Review of English Studies*, volume 30, pages 308-316.

Wilson, J.F., Weiss, D.A., Richards, M., Thomas, M.G., Bradman, N. and Goldstein, D.B. (2001) Genetic evidence for different male and female roles during cultural transitions in the British Isles. *Proceedings of the National Academy of Science*, USA, volume 98, pages 5078-5083.

Wood, E.T., Stover, D.A. and Ehret, C., et al, (11 co-authors) (2005). Contrasting patterns of Y chromosome and mtDNA variation in Africa: evidence for sex-biased demographic processes. *European Journal of Human Genetics*, volume 13, pages 867-876.

Wood, M. (2000) *In Search of England; Journeys into the English Past*, Penguin, London.

Wood, M. (2010) *The Story of England*, Penguin, London.

Y Chromosome Consortium (2002) A nomenclature system for the tree of human Y-chromosomal binary haplogroups. *Genome Research*, volume 12, pages 339-348.

Index

Other related books

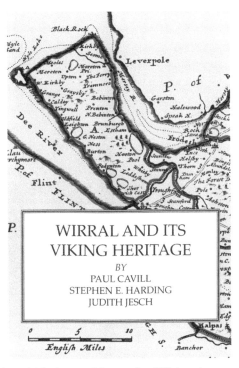

This book, first published in 2000, is a guide to the Viking impact on the Wirral. It includes reprinted illustrated essays from F. T. Wainwright, John McN. Dodgson, J. D. Bu'Lock and W. G. Collingwood, on the history, art and names of the region. And the work is brought up to date by original contributions from Simon C. Bean, Stephen E. Harding, Judith Jesch and Andrew Wawn on recent developments in the history, archaeology, scholarly and popular interest in the Wirral. It is completed by a gazetteer examining the origins of the major names, which also forms an index to the volume.

The book provides absorbing reading and is an important resource for anyone interested in the past of the Wirral and the origins of its names.

Published by the English Place-Name Society, School of English Studies, University of Nottingham, Nottingham NG7 2RD.
paul.cavill@nottingham.ac.uk
Registered Charity No. 257891. Popular Series, ISBN 0 904889 59 9
Paperback, 235 x 157 mm, ix + 149 pp.

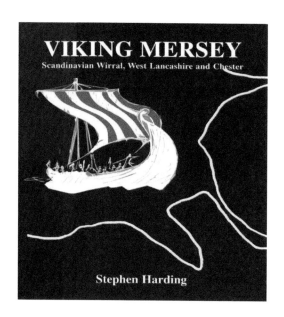

1100 years ago marked the start of a Viking invasion of the Mersey region, which reached out into Chester, West Lancashire and beyond. The Vikings left behind place-names like Kirkby, Kirby, Meols and Croxteth, which can also be found in Iceland, another region they were invading. This book, first published in 2002, is about these people in peace and war, their customs, traditions, pastimes, their paganism and their Christianity, their governments and their financial centre at Chester. It also includes a section on how modern genetic research is being used to discover the descendants of these "Invaders" in the modern day population.

Published by Countyvise Ltd., 14 Appin Road, Birkenhead, CH41 9HH, U.K., cv@birkenheadpress.co.uk
Size 245mm x 210mm paperback, 240 pages, fully illustrated. ISBN 9781901231342

2nd edition (2006) of the highly successful book first published in 2000. 1100 years ago a group of Viking settlers from Norway arrived somewhere between Vestri-Kirkjubyr (West Kirby) and Melr (Meols) on the shores of north Wirral - a small peninsula lying between the Rivers Dee and Mersey - having been driven out of Ireland. This initiated a mass migration of their fellow countrymen into the area and soon they had established a community with a clearly defined border, its own leader, its own language, a trading port, and at its centre a place of assembly or government - the Thing at Þingvollr (Thingwall). This community was answerable to nobody else: the English, the Welsh, the Dublin Norse, the Isle of Man, Iceland, and not even Norway. The Wirral Norse settlement therefore satisfied all the criteria of an independent, self-governing Viking State - albeit a mini one! This book, written by Wirral-exile and scientist Steve Harding, is about these people, why they left Norway, where they settled, their religion, their pastimes, and the legends that have been attributed to them by the Victorians. Wirral was also probably witness to one of the greatest battles in the history of the British Isles - Brunanburh.

Published by Countyvise Ltd., 14 Appin Road, Birkenhead, CH41 9HH, U.K., cv@birkenheadpress.co.uk
Size 220mm x 155mm hardback, 240 pages, fully illustrated. ISBN 978901231601

T - #1024 - 101024 - C166 - 246/208/9 - PB - 9781466590854 - Gloss Lamination